Plant Life and Environment Dynamics

This book series aims to publish volumes focusing on plant responses to changing environmental conditions. The series is cenrted on the theme of plant-environmnet interaction, ecosystem changes, plant physiology, various physio-biochemical and molecular mechanism of adpatation wokring within the plants. The book series also provides knowledge and understanding of plant's stess tolerance strategies such as ionomic transcriptomic, proteomic, phytohaormoes, cellular redox regulation, energy mentabolism, nanomateiral in plant protection and genomics of plants under environmnetal stress metabolomic approaches which are currently being used in order to protect plant life against adverse effects of changing environment. This series will cover book ideas focusing on all aspects of plant's responses to the change in the respective environment. The books published in this series are a useful source of information for academicians and researchers working in the field of agriculture, botany, agronomy, biotechnology, plant science, crop physiology, plant stress physiology, plant and environment interactions related courses.

Series plans to cover dynamics of plant-environment interaction via following broad themes:

(a) Abiotic stress and plant development
(b) Photosynthesis and energy metabolism
(c) Bio-stimulants and plant life
(d) Phytohormones in stress tolerance
(e) Cellular redox regulation (ROS, RNS and RSS), oxidative stress and antioxidant defense system
(f) Transcriptomics, Proteomics and Functional Genomics
(g) Ionomics: Uptake and Assimilation
(h) Nanomaterials and plant life

Anket Sharma • Golam Jalal Ahammed
Editors

Melatonin in Plants: Role in Plant Growth, Development, and Stress Response

Springer

Editors
Anket Sharma
Institute of Genomics for Crop Abiotic Stress Tolerance
Texas Tech University
Lubbock, TX, USA

Golam Jalal Ahammed
Department of Horticulture
Henan University of Science and Technology
Luoyang, Henan, China

ISSN 2730-6755 ISSN 2730-6763 (electronic)
Plant Life and Environment Dynamics
ISBN 978-981-99-8050-5 ISBN 978-981-99-8051-2 (eBook)
https://doi.org/10.1007/978-981-99-8051-2

This Springer imprint is published by the registered company Springer Nature Singapore Pte Ltd.
The registered company address is: 152 Beach Road, #21-01/04 Gateway East, Singapore 189721, Singapore

This book is dedicated to the phytomelatonin research community for advancing melatonin research in plants across the world.

Preface

Plants are continuously exposed to biotic and abiotic stress conditions, which have negative impacts on plant growth and development. Stress factors disturb normal biological functions resulting in a significant decrease in yield. However, the innate defense system of plants attempts to counterattack the damaging effects of adverse environmental conditions by modulating various physiological and metabolic processes. But, beyond a threshold level of a specific stress factor, the plant defense system becomes ineffective to protect plants from oxidative damage. However, the exogenous application of plant hormones and growth regulators can help plants to withstand abiotic stresses, which is considered one of the best eco-friendly approaches to improve plant resistance against biotic and abiotic stresses. Currently, a lot of investigations are focused on examining the exact mechanisms behind hormone-mediated regulation of plant growth under environmental stresses.

Melatonin is an important biologically active compound that acts as a multifunctional molecule and regulates key physiological and biochemical processes in plants. Melatonin is well known to regulate redox homeostasis in plants under stress conditions and helps in protecting plant cells from the damaging effects of excessive reactive oxygen species. Currently, researchers all over the globe have been exploring the in-depth mechanisms of melatonin-modulated regulation of plant biology using various advanced molecular techniques. These recent advancements in melatonin research can be beneficial to explore some important biological pathways, which have possible applications in the future for developing stress-tolerant crop varieties. Keeping all these facts in mind, we aimed to compile the current book to highlight recent developments in the arena of melatonin-mediated regulation of abiotic and biotic stress.

This book comprises 13 chapters. Chapter 1 focuses on the history, biosynthesis, and functions of phytomelatonin. Chapter 2 deals with the role of melatonin in plant growth and development. Chapters 3–5 explain the regulation of photosynthesis and the crosstalk of melatonin with phytohormones and other signaling molecules. Chapters 6–10 explain the role of melatonin in plants under various abiotic stresses such as high light, UV radiations, drought, salt, cold, and heavy metals. These chapters provide an in-depth literature review of the physiological and molecular responses regulated by melatonin in plants under abiotic stress. In Chap. 11, the authors discuss melatonin-mediated regulation of plant biology under biotic stress. Chapter 12 is dedicated to the roles of melatonin in the post-harvest management of various crops. The final Chap. 13 deals with recent developments in the quantification of melatonin content in plants.

Texas, TX, USA Anket Sharma
Luoyang, Henan, China Golam Jalal Ahammed

Acknowledgments

This work was supported by the National Natural Science Foundation of China (grant number 31950410555); Ministry of Science and Technology of the People's Republic of China (grant numbers DL2022026004L, QNJ2021026001, QNJ20200226001); Henan International Joint Laboratory of Stress Resistance Regulation and Safe Production of Protected Vegetables; Innovative Research Team (Science and Technology) in University of Henan Province (Grant number 23IRTSTHN024); and Henan University of Science and Technology Research Start-up Funds for New Faculty (grant numbers 13480058, 13480070).

Contents

Editors and Contributors

About the Editors

Anket Sharma currently works as Visiting Scientist at Texas Tech University, Lubbock, USA. Dr. Sharma is an Associate Professor (Adjunct) at Zhejiang A&F University, China. Before joining as a faculty, he worked as a postdoctoral scientist at Zhejiang A&F University, China. He did his Ph.D. in Plant Science from Guru Nanak Dev University, India. He also got a DST-INSPIRE fellowship from the Department of Science & Technology, Government of India. His research interests are to study the plant hormone-mediated regulation of plant physiology under abiotic stress conditions. He has authored over 100 publications in internationally reputed journals. Dr. Sharma has been listed in the "Highly Cited Researchers 2022" by Clarivate Analytics .

Golam Jalal Ahammed is an Associate Professor at the Department of Horticulture, College of Horticulture and Plant Protection, Henan University of Science and Technology, Luoyang, China. Dr. Ahammed received a Ph.D. in Olericulture with a major focus on Plant Stress Physiology and Hormonal Regulation in 2012 from Zhejiang University, China. Afterward, he completed two consecutive postdoctoral programs at the Institute of Pesticide and Environmental Toxicology and the Institute of Crop Science of Zhejiang University. His major research interests include plant stress physiology, phytohormones, climate change effect on plants, biological control of plant diseases, and environmental pollution. Dr. Ahammed authored over 160 peer-reviewed papers in internationally reputed journals. Dr. Ahammed has been recognized on the annual Highly Cited Researchers 2023 list from Clarivate. He is a Senior Editorial Board Member of *Scientific Reports* and an Associate Editor of *Journal of Plant Growth Regulation*, *AoB Plants*, and *BMC Plant Biology* .

Contributors

Golam Jalal Ahammed Department of Horticulture, Henan University of Science and Technology, Luoyang, Henan, China

S. Maqbool Ahmed Botany Section, School of Sciences, Maulana Azad National Urdu University, Hyderabad, Telangana, India

Vasudha Bansal Department of Foods and Nutrition, Government Home Science College, Chandigarh, India

Savita Bhardwaj Department of Botany, MCM DAV College, Kangra, Himachal Pradesh, India
Department of Botany, School of Bioengineering and Biosciences, Lovely Professional University, Phagwara, Punjab, India

Shuangchen Chen College of Horticulture and Plant Protection, Henan University of Science and Technology, Luoyang, P.R. China

Jinxia Cui Department of Horticulture, College of Agriculture, Shihezi University, Shihezi, China

Mohammad Faizan Botany Section, School of Sciences, Maulana Azad National Urdu University, Hyderabad, Telangana, India

Ahmad Faraz Department of Biotechnology, School of Life Sciences, Glocal University, Saharanpur, India

Monika Garg Department of Agri-Biotechnology, National Agri-Food Biotechnology Institute, Mohali, Punjab, India

Shibei Ge Key Laboratory of Tea Quality and Safety Control, Ministry of Agriculture and Rural Affairs, Tea Research Institute, Chinese Academy of Agricultural Sciences, Hangzhou, PR China

Lakshmi Giri Department of Biotechnology, Chandigarh Group of Colleges, Landran, Punjab, India

Tianmeng Guo College of Horticulture and Plant Protection, Henan University of Science and Technology, Luoyang, P.R. China

Yanliang Guo State Key Laboratory of Crop Stress Biology for Arid Areas, College of Horticulture, Northwest A&F University, Yangling, Shaanxi, China

Md. Kamrul Hasan Department of Agricultural Chemistry, Sylhet Agricultural University, Sylhet, Bangladesh

Shamsul Hayat Plant Physiology Section, Department of Botany, Aligarh Muslim University, Aligarh, India

Anjuman Hussain Plant Physiology Lab, Department of Botany, Aligarh Muslim University, Aligarh, India

Neetu Jagota Department of Botany and Environment Science, DAV University, Jalandhar, Punjab, India

Dhriti Kapoor School of Biological and Environmental Sciences, Shoolini University, Solan, Himachal Pradesh, India

Harleen Kaur Department of Botany and Environment Science, DAV University, Jalandhar, Punjab, India

Ravneet Kaur Department of Botany and Environment Science, DAV University, Jalandhar, Punjab, India

Rahul Kumar Faculty of Agricultural Sciences, DAV University, Jalandhar, Punjab, India

Hao Li State Key Laboratory of Crop Stress Biology for Arid Areas, College of Horticulture, Northwest A&F University, Yangling, Shaanxi, China

Xin Li Key Laboratory of Tea Quality and Safety Control, Ministry of Agriculture and Rural Affairs, Tea Research Institute, Chinese Academy of Agricultural Sciences, Hangzhou, PR China

Vasudha Maurya Department of Biotechnology, Faculty of Agricultural Sciences, DAV University, Jalandhar, Punjab, India

Maninder Meenu Department of Agri-Biotechnology, National Agri-Food Biotechnology Institute, Mohali, Punjab, India

Vivek S. Mehta Regional Horticulture Research Station, Navsari Agricultural University, Navsari, Gujarat, India

Nazish Department of Science, JMB Institute of Life Sciences and Higher Education, Pilibhit, India

Congrui Qin College of Horticulture and Plant Protection, Henan University of Science and Technology, Luoyang, P.R. China

Ashish Sharma Department of Botany and Environment Science, DAV University, Jalandhar, Punjab, India

Ashutosh Sharma Faculty of Agricultural Sciences, DAV University, Jalandhar, Punjab, India

Indu Sharma Department of Life Sciences, University Institute of Sciences, Sant Baba Bhag Singh University, Jalandhar, Punjab, India

Kanu Priya Sharma Department of Biotechnology, Punjab Agricultural University, Ludhiana, Punjab, India

Neeta Raj Sharma School of Bioengineering and Biosciences, Lovely Professional University, Phagwara, Punjab, India

Keyin Shen Key Laboratory of Tea Quality and Safety Control, Ministry of Agriculture and Rural Affairs, Tea Research Institute, Chinese Academy of Agricultural Sciences, Hangzhou, PR China

Jannat Shopan Key Laboratory of IRFA, Sylhet Agricultural University, Sylhet, Bangladesh

Narayan Singh Department of Biotechnology, DAV University, Jalandhar, Punjab, India

Swapnil Singh Department of Botany and Environment Science, DAV University, Jalandhar, Punjab, India

Anupreet Kaur Sobti Department of Foods and Nutrition, Government Home Science College, Chandigarh, India

Neha Verma Government College of Education, Jalandhar, Punjab, India

Yameng Wang Key Laboratory of Tea Quality and Safety Control, Ministry of Agriculture and Rural Affairs, Tea Research Institute, Chinese Academy of Agricultural Sciences, Hangzhou, PR China

Pei Wu Department of Horticulture, College of Agriculture, Shihezi University, Shihezi, China

Huihang Xu College of Horticulture and Plant Protection, Henan University of Science and Technology, Luoyang, P.R. China

Fangying Zheng College of Horticulture and Plant Protection, Henan University of Science and Technology, Luoyang, P.R. China

Jie Zhou Zhejiang Provincial Key Laboratory of Horticultural Plant Integrative Biology, Department of Horticulture, Zijingang Campus, Zhejiang University, Hangzhou, China

Chapter 1
Phytomelatonin: History, Biosynthesis, and Functions

Maninder Meenu, Mohammad Faizan, Kanu Priya Sharma, Lakshmi Giri, Anupreet Kaur Sobti, Vasudha Bansal, and Monika Garg

Abstract Melatonin synthesized by plants is known as phytomelatonin. It is a ubiquitous indolamine, largely investigated for its key functioning in the regulation of many morpho-physiological activities in plants. The biosynthesis of melatonin in plants starts from tryptophan through many similar successive enzymatic steps. The role of phytomelatonin in many transcriptional processes through altering the genes confirms its significant involvement as a multi-regulatory molecule that leads the many aspects of plant development. The current chapter addresses the history and biosynthesis of phytomelatonin. The potential role of phytomelatonin in agriculture as a growth regulator, photosynthesis, antioxidant defense system, plant hormone interaction, and secondary metabolism are also described briefly.

Keywords Secondary metabolism · Antioxidant defense system · Photosynthetic efficiency · Transcriptional process

M. Meenu (✉) · M. Garg (✉)
Department of Agri-Biotechnology, National Agri-Food Biotechnology Institute, Mohali, Punjab, India

M. Faizan
Botany Section, School of Sciences, Maulana Azad National Urdu University, Hyderabad, Telangana, India

K. P. Sharma
Department of Biochemistry, Punjab Agricultural University, Ludhiana, Punjab, India

L. Giri
Department of Biotechnology, Chandigarh Group of Colleges, Landran, Punjab, India

A. K. Sobti · V. Bansal
Department of Foods and Nutrition, Government Home Science College, Chandigarh, India

A. Sharma, G. J. Ahammed (eds.), *Melatonin in Plants: Role in Plant Growth, Development, and Stress Response*, Plant Life and Environment Dynamics,
https://doi.org/10.1007/978-981-99-8051-2_1

1.1 Introduction

Melatonin (*N*-acetyl-5-methoxytryptamine), one of the most important biochemicals in living organisms, serves as a secondary metabolite associated with a variety of physiological, hormonal, and natural processes at the cell, tissue, and organ levels. The tryptophan metabolic pathway produces the hormone melatonin as its by-product. Studies have revealed that it is still a part of an evolutionary process and that its origins may be linked to prokaryotes (Manchester et al. 2015; Zhao et al. 2021a). In animals, as a chronobiological hormone that alerts the brain and other peripheral organs to darkness, melatonin is particularly significant. It regulates sleep-wake cycles and synchronizes life activity into seasonal periods and reproductive activities. It is an endogenous synchronizer for both endocrine via neurotransmitter release and other physiological rhythms (Lerner et al. 1959).

Melatonin of plant origin is known as phytomelatonin (*N*-acetyl-5-methoxytryptamine). Phytomelatonin has been found in almost all plant species examined since its discovery in 1995. However, the actual quantity of phytomelatonin significantly varies among plant species. Typically, the phytomelatonin content in aromatic medicinal plants is higher compared to other plants. The conditions for growth and development are the major factors, which exhibit the potent impact on endogenous phytomelatonin content. Plant cellular organelles such as chloroplasts and mitochondria are involved in the synthesis of phytomelatonin. The presence of phytomelatonin synthesizing enzymes in cytoplasm and endoplasmic reticulum indicates the involvement of these organelles in melatonin production. However, it is still unknown if the production of melatonin in various cellular compartments may be influenced by plant species, types of stress, and serotonin levels. Phytomelatonin is also present in the vascular tissue, but its role in transportation and assimilation of plant nutrients still needs to be explored. Till now, plant roots have not been explored for the presence of melatonin.

The presence of high amount of melatonin in plants triggered plant scientists to explore its regulatory functions in plant growth and development. Phytomelatonin is mentioned as a pleiotropic hormone that exhibit a wide range of advantageous effects that generally enhance physiological processes such as seed germination and growth, photosynthesis including pigment content, photorespiration, stomatal conductance, and water economy, seed and fruit yield, osmoregulation, and control of various metabolic pathways such as carbohydrates, lipids, nitrogen compounds, sulfur, and phosphorus cycles. Phytomelatonin stimulates the formation of simple phenols, flavonoids, anthocyanins, carotenoids, and numerous terpenoids in relation to secondary metabolism. Phytomelatonin delays leaf senescence and encourages rooting processes. Melatonin is also reported to control ethylene levels and lycopene content in fruits during post-harvest handling. Thus, phytomelatonin exhibit significant impact on overall metabolism of ripening of agri-produce. Melatonin is also reported to induce parthenocarpy during fruiting by regulating genes involved in gibberellin pathways. Phytomelatonin also aids in crop health by controlling damage caused due to pathogen infections by promoting systemic acquired resistance.

Several physiological investigations also have mentioned phytomelatonin as a defensive biomolecule that can efficiently resist various kinds of physical, chemical, and biological stresses. Phytomelatonin is also a potent plant hormone regulator which exhibit a significant impact on auxin, gibberellins, cytokinins, abscisic acid, ethylene, jasmonic acid, salicylic acid, and brassinosteroids. Thus, melatonin is frequently known as a "plant master regulator." All these above-mentioned effects of phytomelatonin are concentration-dependent, thus it is necessary to adopt effective approaches for melatonin detection in plant samples.

The role of phytomelatonin in controlling plant-ecosystem interactions has also been highlighted by the characteristics of melatonin in phytoremediation, plant innate immunity, and plant-rhizomicrobial population.

1.2 History of Phytomelatonin Discovery and Major Findings

1.2.1 Melatonin Discovery in Plants (Phytomelatonin)

As mentioned in Sect. 1.1, till 1995, melatonin was only recognized as an animal hormone. But in 1993, van Tassel and O'Neill found melatonin in *Pharbitis nil* L by employing radioimmunoassay (RIA) and gas chromatography (GC) equipped with mass spectrometry (MS). But this finding was not released until 1995 (Arnao 2014). Interestingly, in 1995 melatonin levels were measured in *Nicotiana tabacum* L. and five culinary plants was published by Dubbels and colleagues using RIA and HPLC-MS (Dubbels et al. 1995). Two months later, a new study measured melatonin in numerous plant food extracts using liquid chromatography (HPLC) and fluorescence detection was published (Hattori et al. 1995). The same year, another article was published on the presence of melatonin in *Chenopodium rubrum* L. by Czech researchers (Kolar et al. 1995). Followed by these findings, multiple investigations that have quantified melatonin in different plants which established the present of this chemical in both plants and mammals. After the discovery of plant melatonin, another study on metabolic processes similar to those associated with animal melatonin was published. The first goal of this study was to show how it might have a role as a chrono-regulatory protein in photomorphogenic processes like flowering. The Czech researchers Kolá and Macháková devoted many years to explore the impact of phytomelatonin on *Chenopodium rubrum* L. flowering and how circadian rhythms affects the phytomelatonin levels, but their attempts were unsuccessful (Kolar et al. 1997, 2003; Wolf et al. 2001; Kolar 2003). In 2000, another research by a Canadian team led by Dr. Saxena on St. John's wort (*Hypericum perforatum* L.) cell culture discovered the stages in phytomelatonin biosynthesis pathway that were comparable to those of mammals. It was suggested that melatonin might function as an auxin in vitro cell culture as melatonin and indole-3-acetic acid (IAA) are chemically comparable (Murch et al. 2000, 2001;

Murch and Saxena 2002). In 2004, a study was published to demonstrate growth-stimulating impact of melatonin on etiolated hypocotyls of lupin (*Lupinus albus* L.). The stimulatory potential of melatonin was reported to be 63% higher compared to the IAA (Hernández-Ruiz et al. 2004). This stimulatory potential of melatonin was later verified in several Poaceae species (Hernández-Ruiz et al. 2005; Arnao and Hernández-Ruiz 2006).

In 2004, another study revealed the protective effects of melatonin in carrot (*Daucus carota* L.) cells exposed to cold stress (Lei et al. 2004). Since 2008, several studies reported the protective effect of melatonin against different kinds of stress such as the impact of melatonin on the germination of *Brassica oleracea* L. seeds under copper toxicity (Posmyk et al. 2008, 2009). and protective effect of melatonin against abiotic (drought, salinity, waterlogging, cold, heat, metal toxins, herbicides, UV radiation) and biotic (bacteria, fungi, virus) stressors (Arnao and Hernández-Ruiz 2014, 2015, 2019; Debnath et al. 2019; Zhan et al. 2019; Kul et al. 2019; Wang et al. 2018; Fan et al. 2018; Sharif et al. 2018; Kanwar et al. 2018; Yu et al. 2018; Sharma and Zheng 2019; Sharma et al. 2020). All these stressors causes a spike in reactive oxygen and nitrogen species (ROS/RNS), which in turn triggers the biosynthesis of phytomelatonin in tissues. The enzymes involved in the biosynthesis of melatonin such as tryptophan 5-hydroxylase (T5H), tryptophan decarboxylase (TDC), serotonin *N*-acetyltransferase (SNAT), acetylserotonin methyltransferase (ASMT), and caffeic acid *O*-methyltransferase (COMT) exhibit increased gene expression in response to induced stress (Arnao and Hernández-Ruiz 2018, 2019; Hernández-Ruiz and Arnao 2018). In 2007, melatonin was reported to induce the formation of adventitious roots in lupin by stimulating the formation of root primordia in the pericycle (Arnao and Hernández-Ruiz 2007).

A study published in 2009 revealed the delay in chlorophylls loss due to darkness in barley plants followed by melatonin treatment (Arnao and Hernández-Ruiz 2009a, b). Researchers conducted in later years revealed the control of transcription factors connected to senescence and chlorophyll biosynthesis in leaf senescence by melatonin (Wang et al. 2012).

Overall, only 35 articles related to phytomelatonin were published between 1995 and 2005. Whereas, the number of publications (264) increased almost linearly between 2009 and 2017, and an exponential rise in publications (388) was observed from 2018 to 2020. Whereas, in year 2021, there were 287 articles published, that number was rose to 372 articles in 2022. Till March 2023, 98 publications related to phytomelatonin are already available online. The number of publication related to phytomelatonin is expected to be increased significantly in near future due to the significant role of melatonin in plant species.

1.2.1.1 Major Studies and Breakthroughs Related to Phytomelatonin

Phytomelatonin is a superior antioxidant with a much higher protective capacity compared to the traditional antioxidants such as ascorbic acid and vitamin E. Its ability to control gene expression quickly led to the development of novel theories.

For a molecule to be classified as a plant hormone, it is important to know its production, degradation, potential conjugation, transportation, receptors, signaling pathways and physiological effects. In case of phytomelatonin, many of these parameters are well-documented except identification of its receptor which is the most challenging task to deal with (Arnao and Hernández-Ruiz 2018, 2019). In 2018, Dr. Chen and co-workers discovered phytomelatonin receptor known as PMTR1 in plasma membrane of *Arabidopsis thaliana*. Phytomelatonin can increase its expression in various tissues, as it works with the G-protein subunit (GPA1). The dissociation of G-protein subunits Gγb and Gα is caused by PMTR1-phytomelatonin binding initiates an NADPH oxidase-dependent H_2O_2 generation (RBOH) that increases Ca^{2+} inflow and encourages K^+ efflux which in turn lead to stomatal closing (Wei et al. 2018). This was the most important finding which established phytomelatonin as a hormone. However, it is also important to describe the phytomelatonin receptor in other plant species. Phytomelatonin was known for its regulatory functions in the redox network, where it stimulates the activation of many genetic resources during the times of stress conditions to rebalance the redox homeostatsis. Phytomelatonin was also mentioned as a major plant regulator of the redox network (Arnao and Hernández-Ruiz 2019; Hernández-Ruiz and Arnao 2018; Garrido et al. 2010). Phytomelatonin has also been shown to act as a chronoregulator in plants (Arnao and Hernández-Ruiz 2020), but the experimental confirmation for the relationship between biological clock and redox network is still pending. If this hypothesis is true, phytomelatonin (a plant master regulator) proves to functions similar to the animal melatonin in a way it transfers biological clock oscillations to the redox network which provides an appropriate response to achieve redox homeostasis in stressful conditions. Since 2010, different parts of various wild and edible plants as well as plant-based cooked and fermented food products (juices, musts, wines, and infusions) have been explored for the presence of phytomelatonin (Arnao and Hernández-Ruiz 2020).

Phytomelatonin mentioned to exhibit several physiological functions (Arnao and Hernández-Ruiz 2020) such as regulation of lipids and carbohydrates metabolism, regulation of nitrogen, sulfur, and phosphorus cycles, stimulation of secondary metabolism to produce carotenoids, anthocyanins, and flavonoids under stress. Melatonin causes parthenocarpy in pears, which raises the amount of gibberellins (Liu et al. 2018). In Arabidopsis, melatonin levels are controlled by diurnal rhythms and are linked to Circadian Clock Associated 1, gene that regulates circadian rhythms. Additionally, exogenous melatonin and circadian variations co-regulated the expression of AtCBFs and C-repeat-binding factors (CBFs/DREB1s), suggesting a potential connection between the clock, endogenous melatonin level, and plant defense (Shi et al. 2016). Phytomelatonin also controls its own production by increasing the expression of involved genes such as SNAT (serotonin *N*-acetyltransferase), ASMT (acetylserotonin *O*-methyltransferase), and COMT (caffeic acid 3-*O*-methyltransferase) (Mannino et al. 2021). Phytomelatonin controls the production of genes associated with and triggered by foliar senescence as well as genes linked to chlorophyll degradation (Hernández-Ruiz et al. 2021). In addition, during the rooting processes of primary, secondary, and adventitious roots, phytomelatonin controls the production of several factors such as PIN auxin transporters and AUX1 (Arnao and Hernández-Ruiz 2018). Phytomelatonin mentioned to improve plant health by slowing the harm and promoting systemic acquired resistance (SAR) in response to bacterial, fungal, and viral pathogen infections (Zeng et al. 2022a). Phytomelatonin was also known to control several enzymes in the cell wall, ethylene biosynthesis, and primary and secondary metabolisms. In addition, phytomelatonin enhanced the ethylene and lycopene concentration in fruits during post-harvest regulation along with maintaining the freshness of cut flowers (Xu et al. 2019).

1.3 Biosynthesis of Phytomelatonin

The chloroplast and mitochondria are hypothesized to be the major sites involved in the biosynthesis of melatonin since these cellular compartments are reported to have the highest melatonin levels in plants. Moreover, the localization of serotonin *N*-acetyltransferase (SNAT) in these organelles supports the above hypothesis. The serotonin *N*-acetyltransferase is a rate-limiting enzyme in the biosynthesis of melatonin. Phytomelatonin is synthesized from amino acid tryptophan. Plants can synthesize tryptophan de novo by the shikimate pathway. There are six enzymes known to be involved in the synthesis of melatonin (1) L-tryptophan decarboxylase (TDC), (2) tryptamine 5-hydroxylase (T5H), (3) serotonin *N*-acetyltransferase (SNAT), (4) acetylserotonin *O*-methyltransferase (ASMT), (5) caffeic acid 3-*O*-methyltransferase (COMT), and (6) a putative tryptophan hydroxylase (TPH) not yet identified. TDC, ASMT, and COMT are localized in the cytoplasm, while T5H lies in the endoplasmic reticulum and SNAT is expressed in chloroplasts (Back 2021; Zhao et al. 2021b).

The first step in melatonin biosynthesis is the conversion of tryptophan into serotonin. This conversion may occur via two different pathways (Back et al. 2016). Tryptophan is transformed into serotonin as the initial step in the production of melatonin. There are two different ways through which this conversion can take place. One method involves the decarboxylation of tryptophan to produce tryptamine. This step is catalyzed by L-tryptophan decarboxylase. Tryptamine is then converted into serotonin by the tryptamine 5- hydroxylase. Otherwise, the hydroxylation of tryptophan into 5-hydrotryptophan is followed by decarboxylation into serotonin. The above steps were catalyzed by tryptophan hydroxylase and L-tryptophan decarboxylase, respectively. Since, tryptophan decarboxylase has an affinity for both tryptophan as well as 5-hydoxytryptophan, both above-mentioned pathways are possible.

In plants, decarboxylation has been demonstrated as the first step followed by hydroxylation (Back et al. 2016). Furthermore, the synthesis of melatonin from serotonin is two-step pathway that involves three distinct enzymes, that is, serotonin *N*-acetyltransferase (SNAT), *N*-acetylserotonin methyltransferase (ASMT), and caffeic acid *O*-methyltransferase (COMT). SNAT catalyzes the reaction that changes

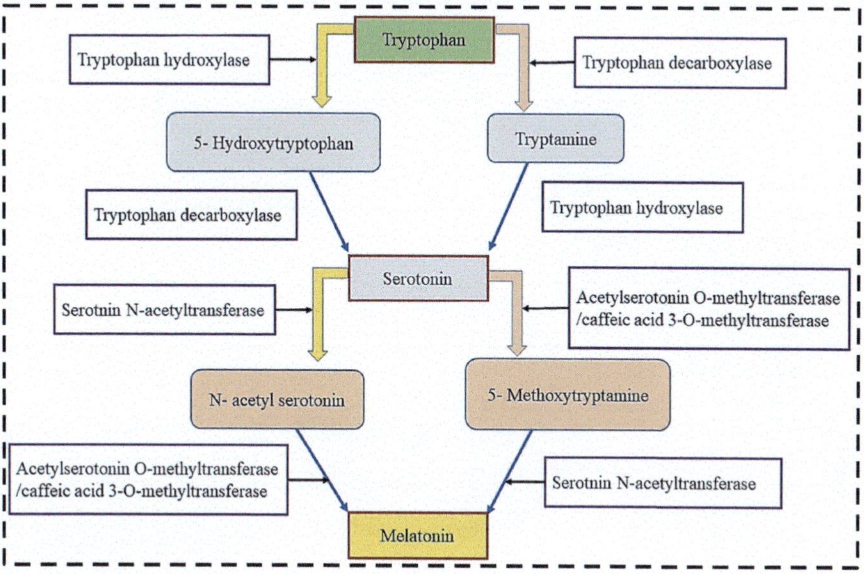

Fig. 1.1 Biosynthesis of melatonin in plants

serotonin into *N*-acetylserotonin, which is then converted into melatonin by either ASMT or COMT. Because SNAT has a substrate affinity for serotonin and 5-methoxytryptamine (SNAT), the process produces *N*-acetylserotonin, which is then *O*-methylated to melatonin by ASMT/COMT. Similarly, ASMT and COMT show substrate affinity for serotonin and N-acetylserotonin, with ASMT and COMT first methylating serotonin to 5-methoxytryptamine, followed by SNAT reaction to melatonin. In conclusion, SNAT and ASMT/COMT can catalyze the conversion of serotonin into *N*-acetylserotonin and 5-methoxytryptamine, which is followed by the creation of melatonin by ASMT/COMT and SNAT, respectively (Zhao et al. 2021b; Back et al. 2016) (Fig. 1.1).

1.4 Functions of Phytomelatonin

Phytomelatonin exhibit several important functions to improve growth and yield of plants such as improvement in photosynthetic activity, antioxidant defense system, regulates redox network, primary and secondary metabolism, defense against apoptosis, and stressful conditions as shown in Fig. 1.2. In the following section, the function of phytomelatonin in plant growth and yield are discussed in detail.

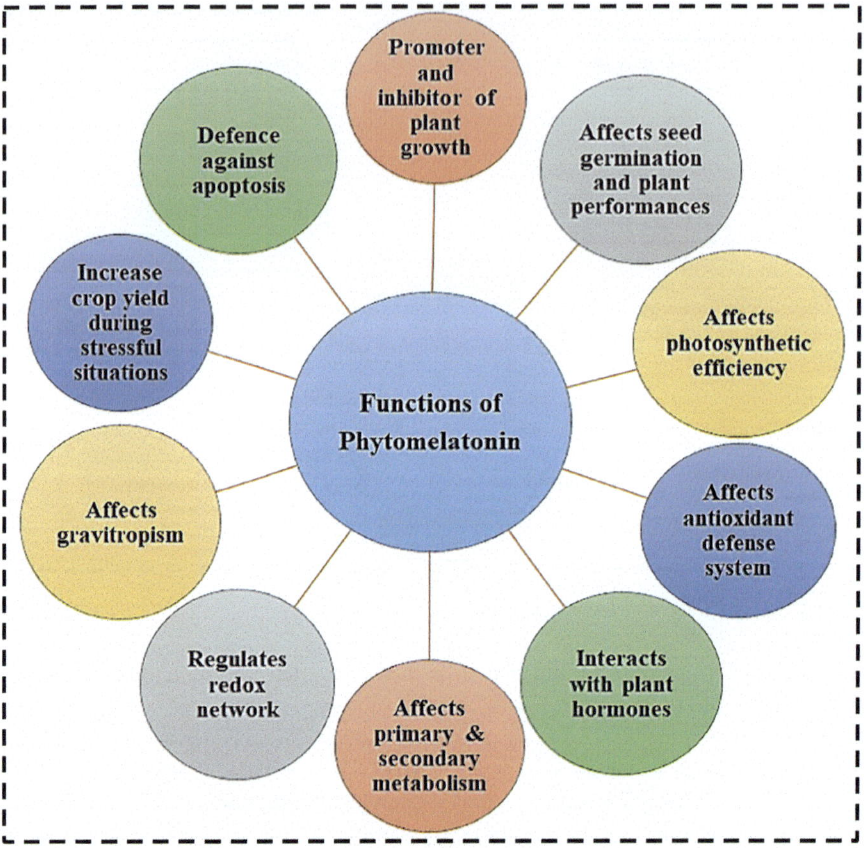

Fig. 1.2 Different functions of phytomelatonin to enhance plant growth and yield

1.4.1 Impact of Phytomelatonin on Plant Growth

Following by the initial discovery of melatonin in plants, researchers have also linked it to the regulation of growth and development of variety of plants such as *Arabidopsis thaliana*, oat, mustard, cauliflower, cucumber, sunflower, barley, lupin, rice, Prunus, guava, soybean, tomato, wheat and corn plants (Gao et al. 2015; Arnao and Hernández-Ruiz 2017, 2018). Phytomelatonin is structurally quite similar to the IAA (Indole-3 Acetic Acid), a commonly occurring plant hormone of auxin class. According to a previous research, phytomelatonin and IAA might collaborate in several physiological processes relating to plant growth and development as their similar structures may serve related purposes. Both phytomelatonin and IAA also have the ability to act as antioxidants (Li et al. 2016). In St. John's wort in vitro cultures, an increase in endogenous melatonin concentration considerably stimulated the root growth, whereas the build-up of serotonin, a precursor to melatonin, directs shoot production (Arnao and Hernández-Ruiz 2007).

Murch et al. (2009) presented the first encouraging study on the regulatory mechanism of melatonin in relation to auxin during plant growth and development. Researchers have identified potential interactions between melatonin and auxin using metabolic inhibitors of auxins, serotonin, and melatonin. The finding of this study emphasized on the possibility that melatonin and auxin interact to influence root architecture, hence promoting plant growth and development.

Till now, only a few genes and metabolites have been reported to be involved in the synthesis and degradation of phytomelatonin, which limit the understanding of melatonin-auxin-mediated interactions in plants. Thus, significant research should be conducted to identify the additional crucial elements involved in the regulation of phytomelatonin production and signaling. In addition, identification and characterization of novel gene(s) associated with phytomelatonin production and signaling will offer novel insights of the associated processes. It is also important to employ effective instruments and sophisticated methodologies for melatonin detection in plant samples.

1.4.2 Impact of Phytomelatonin on Seed Germination and Plant Performance

Seed germination is the most crucial stage of crop growth. The genetic makeup of a crop as well as its surroundings have a significant impact on its germination. Water is the one of these traits that has the greatest environmental impact on seed germination. Recent research has demonstrated that crop plants may experience severe hypoxic stress due to prolonged flooding or submersion, which will significantly reduce the crop yield. Phytohormones are among the main elements which regulates the seed germination. Studies have reported the activation of germination-related genes in the presence of plant hormones. Hence, phytohormone therapy is successfully employed to enhance seed germination. The widespread distribution of melatonin across fungi, algae, bacteria, mammals, and plants has led to the hypothesis that it may function as a photoperiodic and circadian rhythm regulator as well as a universal antioxidant (Paredes et al. 2009; Tan et al. 2012). Melatonin and a variety of its metabolites are known as endogenous free radical scavengers and powerful broad-spectrum antioxidants and reported to directly scavenge H_2O_2, which in turn maintain intracellular H_2O_2 concentrations (Reiter et al. 2007).

Previously, Korkmaz et al. (2017) found that melatonin treatment significantly reduced the H_2O_2 levels and MDA accumulation due to increased antioxidant enzyme activity and direct free radical scavenging by melatonin. In another study, Posmyk et al. (2009) showed that spraying melatonin on cucumber seed provides considerable protection to their membrane structures against peroxidation and MDA accumulation under cold stress. In this study, melatonin was reported to enhance the activity of antioxidant enzymes and demonstrate its critical role in providing antioxidant defense under cold stress circumstances. Cold-stressed pepper seeds and seedlings showed increased activity of SOD, CAT, and POX followed by melatonin treatment which revealed the involvement of melatonin in free radical scavenging. Similar findings were also reported by Zhang et al. (2013), who found that melatonin treatment significantly increased the activity of antioxidant free radical scavenging enzymes such as SOD, CAT, and POX in cucumber seeds germinating under drought stress conditions.

Zeng et al. (2022b) reported a significant increase in the seed germination, seedling growth, and early stand establishment of rice seed soaked in melatonin. Compared to the control, seeds soaked in melatonin presented a significant increase in the growth of rice seed shoots and roots, superoxide dismutase (SOD) activity, peroxidase (POD) activity (Posmyk and Janas 2009), and lower the content of malondialdehyde (MDA) which ameliorate the inhibitory effect of flooding stress on rice growth. Overall, soaking rice seeds in 100 μM melatonin would be a reasonable choice for bio stimulation. Melatonin treatment will be helpful in faster germination and survival of rice seedlings under flooding stress. Melatonin can also be used as seed coating to enhance the crop yield and their ability to withstand abiotic stress. An extensive research is required to explore the possibility of melatonin to be used as commercial seed coating treatment.

1.4.3 Impact of Phytomelatonin on Photosynthetic Efficiency of Plants

An essential plant metabolic process called Photosynthesis is an important metabolic process responsible to enhance crop output and carbon uptake. Cold stress is reported to significantly reduce the photosynthetic efficiency of plants. Melatonin, a novel plant growth regulator, can significantly regulates a variety of abiotic stress responses in plants. The molecular basis of melatonin-mediated photosynthetic control in plants under cold stress is not fully understood. According to a study by Altaf et al. (2022), spraying pepper seedlings with melatonin (200 μM) during low-temperature stress (15/5 °C for 7 days) improved gas exchange properties, amount of photosynthetic pigment and expression of their biosynthetic genes. Under cold stress conditions, photochemical activity of photosystems II (PSII) and I (PSI), specifically their maximum quantum efficiencies PSII (Fv/Fm) and PSI, was considerably reduced. In cold stress conditions, melatonin reported to significantly enhance the activity of photosynthetic enzymes (rubisco and fructose-1, 6-bisphosphatase), and quantity of starch, sucrose, soluble sugar, and glucose. In contrast, melatonin treatment significantly enhanced the photochemical activity of PSII and PSI. Melatonin treatment significantly reduced the effects of cold stress, which also resulted in significant reductions in actual PSII efficiency (ΦPSII), electron transport rate (ETR), and photochemical quenching coefficient (qP), while improving nonphotochemical quenching (NPQ). Overall, melatonin treatment significantly reduced the negative impacts of cold stress by maintaining photosynthetic efficiency, which may be useful when genotypes for CS tolerance are being explored.

Due to overuse of fertiliser and other anthropogenic activities, agricultural soil contains a significant amount of hazardous metals, which presented an adverse effect on crop plants, their yield and quality. The application of melatonin by Ayyaz et al. (2020) mentioned to enhance plant tolerance toward environmental stress and also increased plants growth and development under stress condition. Chromium stress mentioned to reduce energy trapping effectiveness by degrading oxygen evolving complex, however, the exogenous application of melatonin protects the oxygen evolving complex of PSII and assisted in maintaining PSII activity. Due to the beneficial effects of melatonin in plants, it is anticipated that in future it may play an important role in developing transgenic crops that exhibit potent photosynthetic activity under stress conditions.

1.4.4 Impact of Phytomelatonin on Antioxidant Defense System

According to Hardeland and Pandi-Perumal (2005), the kynurenine pathway's direct scavenging of physiological melatonin concentrations by a non-enzymatic contribution and the subsequent activities of the metabolites generated become significant. But at normal levels, melatonin can always have signaling effects. Many antioxidant enzymes are upregulated by melatonin. This has most commonly been shown for glutathione peroxidase and occasionally glutathione reductase, most likely indirectly through GSSG. Occasionally, catalase and the superoxide dismutases Cu, Zn, and/or Mn are elevated in certain tissues. In both mammalian and avian brain, glutathione peroxidase appears to be widely distributed among tissues and is consistently stimulated; upregulations in other organs were less consistent. Melatonin's effects on glutathione metabolism appear to go beyond those already discussed. By providing reducing equivalents (NADPH) for the activity of glutathione reductase and boosting the rate of glutathione synthesis, respectively, stimulation of glucose-6-phosphate dehydrogenase and -glutamylcysteine synthase indirectly enhances the function of glutathione peroxidase. Moreover, melatonin helps prevent the generation of radicals in a variety of separate ways. Prooxidant enzymes are downregulated by it, particularly NO synthases and the 5- and 12-lipoxygenases. In order to restrict the growth in the extremely prooxidant metabolite peroxynitrite and of the free radicals formed from it, namely, •NO2, carbonate (CO3•-), and hydroxyl (•OH) radicals, the commonly documented reduction of NO production is particularly crucial. Limits on inflammatory responses may also be imposed via inhibiting lipoxygenase and NO synthase, while the immunomodulatory effects of melatonin are undoubtedly more complicated and may also involve other effects of melatonin and AMK. Instead of focusing primarily on antioxidant actions for the removal of already created radicals, the effects of melatonin on the respiratory chain give new perspectives for reducing radical generation.

1.4.5 Impact of Phytomelatonin on Plant Hormones Interaction

Plant melatonin appears to be a multi-regulatory chemical with numerous distinct functions in plant physiology, comparable to those seen in animals. Studies on melatonin in plants have become much more prevalent in recent years. Melatonin's effects on biotic and abiotic stress, including those brought on by drought, severe heat, salt, chemical pollution, and UV radiation, among others, are among its most researched effects on plants. Melatonin has a significant role in controlling the expression of genes associated to plant hormones, including auxin carrier proteins, indole-3-acetic acid (IAA), gibberellins, cytokinins, abscisic acid, and ethylene metabolism. The majority of studies have focused on melatonin's auxin-like action, which works in a manner similar to IAA in that it can induce growth in shoots and roots and encourage the development of new lateral and adventitious roots. In addition to safeguarding photosynthetic systems and associated subcellular structures and processes, melatonin has the ability to delay senescence. Also important is its function as a gene regulator of ethylene-related elements during fruit ripening and post-harvest procedures. Its part in the pathogen–plant interaction is another important consideration. Melatonin, along with other well-known chemicals like nitric oxide and hormones like jasmonic acid and salicylic acid, appears to play a major role in the plant immune response. In this way, the finding of higher melatonin levels in endophytic organisms connected to plants has shed light on a potential novel method of communication between advantageous endophytes and host plants (Arnao and Hernández-Ruiz 2018).

One of the more specific auxin-like functions of melatonin is its ability to promote growth. In the aerial tissues of lupinus, phalaris, triticum, hordeum, arabidopsis, and cucumis, melatonin generated a three to fourfold increase in growth compared with control plants, and a less marked rise in others. Melatonin slows down the induced senescence process as well. It has been shown that melatonin treatment of barley leaves results in a concentration-dependent delay of dark-induced senescence, which reduces chlorophyll loss in detached leaves. The senescence process was also delayed by incubating leaf sections with various amounts of kinetin (a synthetic cytokinin), which resulted in treated leaves losing less chlorophyll and retaining higher levels of total chlorophyll than control leaves. Treatment with melatonin changed the levels of GAs and ABA. Melatonin induced the upregulation of GA biosynthesis genes in cucumber seedlings under salinity, resulting in a high concentration of active GAs like GA3 and GA4, which supported the salt-inhibited germination process. Moreover, melatonin treatment caused the downregulation of 9-cis-epoxycarotenoid dioxygenase (NCED), a crucial enzyme in ABA production, and the overexpression of ABA catabolism genes (two CYP707 monooxygenases), which led to a rapid drop in ABA levels during seed germination under salt stress. Similar findings were made in drought-stricken apple leaves, where melatonin pre-treatment reduced the ABA concentration by half by controlling the same ABA production and catabolism enzymes as previously indicated (Zhang et al. 2014).

The key aim is to identify a potential melatonin receptor in order to better understand the mechanisms that trigger the reactions in plants. Because it is an amphipatic molecule, its dual action as an antioxidant and a gene expression regulator appears obvious, but it still raises questions about its transport within cells (free or protein-assisted), as well as between tissues and organs (through the xylem and phloem), its potential conjugation with carbohydrates or amino acids, the signaling components involved, and its potential sharing with plant hormone mechanisms. Although there are many hints in this work that plant hormones and melatonin are related, there is still much research to be done before any conclusive findings can be made. Future research on melatonin and specific plant hormones, however, will surely shed light on the level of interaction present at the cellular and tissular level as well as any potential synergies or antagonistic interactions in the mediated responses.

1.4.6 *Impact Phytomelatonin on Primary and Secondary Metabolism of Plants*

The data on melatonin's regulatory function in several plant metabolic pathways has come from numerous studies. Its crucial impact on enzyme transcripts and regulatory elements in many organelles (chloroplasts, mitochondria, endoplasmic reticulum), subcellular locations (cytosol, cell wall), and secondary metabolism stands out. Melatonin definitely has an impact on a number of critical metabolic activities, including photosynthesis, the pentose phosphate shunt (Sharkey 2021), gluconeogenesis, glycolysis, the Krebs cycle, and the production of amino acids and fatty acids. Although one of the most studied, there is still much to learn about carbohydrate metabolism. Melatonin appears to be a key player in the fate of carbohydrates made in the chloroplast and cytosol, from the regulation of Rubisco to the processes of glycolysis and fermentation. Melatonin controls the Calvin cycle's generation of triose phosphate, its conversion to hexoses, the amount of starch in the chloroplastic stroma, and the amount of sucrose in the cytosol and cell walls. Melatonin generally promotes the primary metabolism of carbohydrates as well as that of other primary substances including lipids and amino acids. As a result, the metabolic turnover is activated and conditioned to be appropriate for the current physiological circumstances. Melatonin has a variety of regulatory effects. For instance, it affects photosynthesis, increasing the effectiveness of Rubisco and other Calvin cycle-related enzymes, Photosystem I and II, chlorophyll and carotenoid content, and stomatal complex, with the result being a higher net photosynthesis. Additionally, melatonin mobilizes some important pathways in specific carbohydrate metabolism, such as starch and sucrose biosynthesis, through upregulation of SPS.

Melatonin is a critical bio stimulator or plant growth regulator because it regulates numerous aspects of the metabolism of plant hormones, which, in conjunction with the modulation of the redox network, enable the plant to adapt to its environment in a way that minimizes negative effects and increases tolerance to stressors. Melatonin has the power to affect several stages of secondary metabolism, particularly in the phenolic chemical and terpenes biosynthesis, which explains why it can alter carbohydrate metabolism and enhance the levels of sugars in fruits as well as their organoleptic properties. The regulation of anthocyanins and other flavonoids, as well as carotenoids and essential oils, are some of its standout features. Moreover, melatonin treatment has a good impact on crop yield. For example, melatonin treatment has been shown to boost production of the following crops: rapeseed, cucumber, tomato, wheat, rice, and others (Arnao et al. 2021). There are a good number of questions that need to be answered regarding how melatonin affects carbohydrate metabolism, including: how genes in the nucleus, chloroplasts, and mitochondria are regulated; how it interacts with other plant hormones; how it functions in various organs (leaf, stem, root, flowers, fruits); how it affects the accumulation and breakdown of starch in amyloplasts; how it affects the metabolism of sucrose in source and sink tissues; and many. The equilibrium between carbs, fatty acids, and amino acids has yet to be fully understood.

1.5 Conclusion

Melatonin is an important phytohormone and participated in number of developmental activities of the plant. The information contained in this book chapter clarifies the chemical characteristics of phytomelatonin and the typical biosynthetic pathways occurring in plants. The various morpho-physiological functions regulated by phytomelatonin described in a systematic way. Moreover, the history of phytomelatonin was highlighted. In this chapter, plant hormones interaction, secondary metabolism influence, and antioxidant enzymes activity in respect to phytomelatonin give clear edge of this hormone as a beneficial plant growth regulator.

Acknowledgments This work is supported by MK Bhan young researcher fellowship, Department of Biotechnology, New Delhi, India.

References

Altaf MA, Shu H, Hao Y, Mumtaz MA, Lu X, Zhiwei W (2022) Melatonin affects the photosynthetic performance of pepper (*Capsicum annuum* L.) seedlings under cold stress. Antioxidants 11(12):2414. https://doi.org/10.3390/antiox11122414

Arnao MB (2014) Phytomelatonin: discovery, content, and role in plants. Adv Bot 2014:815769. https://doi.org/10.1155/2014/815769

Arnao MB, Hernández-Ruiz J (2006) The physiological function of melatonin in plants. Plant Signal Behav 1:89–95

Arnao MB, Hernández-Ruiz J (2007) Melatonin promotes adventitious- and lateral root regeneration in etiolated hypocotyls of *Lupinus albus* L. J Pineal Res 42(2):147–152. https://doi.org/10.1111/j.1600-079x.2006.00396.x

Arnao MB, Hernández-Ruiz J (2009a) Chemical stress by different agents affects the melatonin content of barley roots. J Pineal Res 46:295–299

Arnao MB, Hernández-Ruiz J (2009b) Protective effect of melatonin against chlorophyll degradation during the senescence of barley leaves. J Pineal Res 46:58–63

Arnao MB, Hernández-Ruiz J (2014) Melatonin: plant growth regulator and/or biostimulator during stress? Trends Plant Sci 19:789–797

Arnao MB, Hernández-Ruiz J (2015) Functions of melatonin in plants: a review. J Pineal Res 59: 133–150

Arnao MB, Hernández-Ruiz J (2017) Growth activity, rooting capacity, and tropism: three auxinic precepts fulfilled by melatonin. Acta Physiol Plant 39(6):1–9. https://doi.org/10.1007/s11738-017-2428-3

Arnao MB, Hernández-Ruiz J (2018) Melatonin and its relationship to plant hormones. Ann Bot 121(2):195–207. https://doi.org/10.1093/aob/mcx114

Arnao MB, Hernández-Ruiz J (2019) Melatonin: a new plant hormone and/or a plant master regulator? Trends Plant Sci 24:38–48

Arnao MB, Hernández-Ruiz J (2020) Is phytomelatonin a new plant hormone? Agronomy 10(1): 95. https://doi.org/10.3390/agronomy10010095

Arnao MB, Hernández-Ruiz J, Cano A, Reiter RJ (2021) Melatonin and carbohydrate metabolism in plant cells. Plan Theory 10(9):1917. https://doi.org/10.3390/plants10091917

Ayyaz A, Amir M, Umer S, Iqbal M, Bano H, Gul HT, Noor Y, Kanwal A, Khalid A, Javed MS, Athar HUR, Zafar ZU, Farooq M (2020) Melatonin induced changes in photosynthetic efficiency as probed by OJIP associated with improved chromium stress tolerance in canola (*Brassica napus* L.). Heliyon 6(7):e04364. https://doi.org/10.1016/j.heliyon.2020.e04364

Back K (2021) Melatonin metabolism, signaling and possible roles in plants. Plant J 105:376–391

Back K, Tan DX, Reiter RJ (2016) Melatonin biosynthesis in plants: multiple pathways catalyze tryptophan to melatonin in the cytoplasm or chloroplasts. J Pineal Res 61:426–437

Debnath B, Islam W, Li M, Sun Y, Lu X, Mitra S, Hussain M, Liu S, Qiu D (2019) Melatonin mediates enhancement of stress tolerance in plants. Int J Mol Sci 20:1040

Dubbels R, Reiter RJ, Klenke E et al (1995) Melatonin in edible plants identified by radioimmunoassay and by high performance liquid chromatography-mass spectrometry. J Pineal Res 18(1):28–31

Fan J, Xie Y, Zhang Z, Chen L (2018) Melatonin: a multifunctional factor in plants. Int J Mol Sci 19:1528

Gao B, Chu J, Yu T, Xu Q, Sun X, Yuan J, Xiong G, Wang G, Wang Y, Li J (2015) Tryptophan-independent auxin biosynthesis contributes to early embryogenesis in *Arabidopsis*. Proc Natl Acad Sci U S A 112(15):4821–4826. https://doi.org/10.1073/pnas.1503998112

Garrido M, Paredes SD, Cubero J, Lozano M, Toribio-Delgado AF, Muñoz JL, Reiter RJ, Barriga C, Rodríguez AB (2010) Jerte valley cherry-enriched diets improve nocturnal rest and increase 6-sulfatoxymelatonin and total antioxidant capacity in the urine of middle-aged and elderly humans. J Gerontol A Biol Sci Med Sci 65:909–914

Hardeland R, Pandi-Perumal SR (2005) Melatonin, a potent agent in antioxidative defense: actions as a natural food constituent, gastrointestinal factor, drug and prodrug. Nutr Metab 2(1):22. https://doi.org/10.1186/1743-7075-2-22

Hattori H, Migitaka MI et al (1995) Identification of melatonin in plants and its effects on plasma melatonin levels and binding to melatonin receptors in vertebrates. Biochem Mol Biol Int 35(3): 627–634

Hernández-Ruiz J, Arnao MB (2018) Relationship of melatonin and salicylic acid in biotic/abiotic plant stress responses. Agronomy 8:33

Hernández-Ruiz J, Cano A, Arnao MB (2004) Melatonin: growth-stimulating compound present in lupin tissues. Planta 220:140–144

Hernández-Ruiz J, Cano A, Arnao MB (2005) Melatonin acts as a growth-stimulating compound in some monocot species. J Pineal Res 39:137–142

Hernández-Ruiz J, Cano A, Arnao MB (2021) A phytomelatonin-rich extract obtained from selected herbs with application as plant growth regulator. Plan Theory 10(10):2143. https://doi.org/10.3390/plants10102143

Kanwar MK, Yu J, Zhou J (2018) Phytomelatonin: recent advances and future prospects. J Pineal Res 65:e12526

Kolar J (2003) Effects of melatonin on circadian rhythms and Photoperiodism in higher plants. PhD Thesis, Faculty of Natural Sciences, Charles University, Prague, Czechia

Kolar J, Machackova I, Illnerova H, Prinsen E, van Dongen W, van Onckelen H (1995) Melatonin in higher plant determined by radioimmunoassay and liquid chromatography-mass spectrometry. Biol Rhythm Res 26:406–409

Kolar J, Machackova I, Eder J, Prinsen E, van Dongen W, van Onckelen H, Illnerova H (1997) Melatonin: occurrence and daily rhythm in *Chenopodium rubrum*. Phytochemistry 44:1407–1413

Kolar J, Johnson C, Machackova I (2003) Exogenously applied melatonin affects flowering of the short-day plant *Chenopodium rubrum*. Physiol Plant 118:605–612

Korkmaz A, Karaca A, Kocacinar F, Cuci Y (2017) The effects of seed treatment with melatonin on germination and emergence performance of pepper seeds. J Agric Sci 23:167–176

Kul R, Esringü A, Dadasoglu E, Sahin U, Turan M, Örs S, Ekinci M, Agar G, Yildirim E (2019) Melatonin: role in increasing plant tolerance in abiotic stress conditions. In: De Oliveira A (ed) Abiotic and biotic stress in plants. London, IntechOpen, pp 1–19

Lei XY, Zhu RY, Zhang GY, Dai YR (2004) Attenuation of cold-induced apoptosis by exogenous melatonin in carrot suspension cells: the possible involvement of polyamines. J Pineal Res 36: 126–131

Lerner A, Case JF, Heinzelman RV (1959) Structure of melatonin. J Am Chem Soc 81(22): 6084–6085. https://doi.org/10.1021/ja01531a060

Li M, Hasan MK, Li C, Ahammed GJ, Xia X, Shi K, Zhou Y, Reiter RJ, Yu J, Xu M, Zhou J (2016) Melatonin mediates selenium-induced tolerance to cadmium stress in tomato plants. J Pineal Res 61(3):291–302. https://doi.org/10.1111/jpi.12346

Liu J, Zhai R, Liu F, Zhao Y, Wang H, Liu L, Yang C, Wang Z, Ma F, Xu L (2018) Melatonin induces parthenocarpy by regulating genes in Gibberellin pathways of 'Starkrimson' Pear (*Pyrus communis* L.). Front Plant Sci 9:946. https://doi.org/10.3389/fpls.2018.00946

Manchester LC, Coto-Montes A, Boga JA, Andersen LB, Zhou Z, Galano A, Vriend J, Tan DX, Reiter RJ (2015) Melatonin: an ancient molecule that makes oxygen metabolically tolerable. J Pineal Res 59(4):403–419. https://doi.org/10.1111/jpi.12267

Mannino G, Pernici C, Serio G, Gentile C, Bertea CM (2021) Melatonin and phytomelatonin: chemistry, biosynthesis, metabolism, distribution and bioactivity in plants and animals—an overview. Int J Mol Sci 22(18):9996. https://doi.org/10.3390/ijms22189996

Murch SJ, Saxena PK (2002) Melatonin: a potential regulator of plant growth and development? In Vitro Cell Dev Biol Plant 38:531–536

Murch SJ, KrishnaRaj S, Saxena PK (2000) Tryptophan is a precursor for melatonin and serotonin biosynthesis in in vitro regenerated St. John's wort (*Hypericum perforatum* L.) plants. Plant Cell Rep 19:698–704

Murch SJ, Campbell SSB, Saxena PK (2001) The role of serotonin and melatonin in plant morphogenesis. Regulation of auxin-induced root organogenesis in *in vitro*-cultured explants of *Hypericum perforatum* L. *in vitro cell.* Dev Biol Plant 37:786–793

Murch SJ, Alan AR, Cao J, Saxena PK (2009) Melatonin and serotonin in flowers and fruits of *Datura metel* L. J Pineal Res 47:277–283

Paredes SD, Korkmaz A, Manchester LC, Tan DX, Reiter RJ (2009) Phytomelatonin: a review. J Exp Bot 60(1):57–69. https://doi.org/10.1093/jxb/ern284

Posmyk MM, Kuran H, Marciniak K, Janas KM (2008) Presowing seed treatment with melatonin protects red cabbage seedlings against toxic copper ion concentrations. J Pineal Res 45:24–31

Posmyk MM, Balabusta M, Wieczorek M, Sliwinska E, Janas KM (2009) Melatonin applied to cucumber (*Cucumis sativus* L.) seeds improves germination during chilling stress. J Pineal Res 46:214–223

Posmyk MM, Janas KM (2009) Melatonin in plants. Acta Physiol Plant 31(1):1–11. https://doi.org/10.1007/s11738-008-0213-z

Reiter RJ, Tan DX, Terron MP, Flores LJ, Czarnocki Z (2007) Melatonin and its metabolites: new findings regarding their production and their radical scavenging actions. Acta Biochim Pol 54(1):1–9

Sharif R, Xie C, Zhang H, Arnao MB, Ali M, Ali Q, Muhammad I, Shalmani A, Nawaz M, Chen P et al (2018) Melatonin and its effects on plant systems. Molecules 23:2352

Sharkey TD (2021) Pentose phosphate pathway reactions in photosynthesizing cells. Cell 10(6): 1547. https://doi.org/10.3390/cells10061547

Sharma A, Zheng B (2019) Melatonin mediated regulation of drought stress: physiological and molecular aspects. Plan Theory 8(7):190

Sharma A, Wang J, Xu D, Tao S, Chong S, Yan D, Zheng B et al (2020) Melatonin regulates the functional components of photosynthesis, antioxidant system, gene expression, and metabolic pathways to induce drought resistance in grafted Carya cathayensis plants. Sci Total Environ 713:136675

Shi H, Wei Y, He C (2016) Melatonin-induced CBF/DREB1s are essential for diurnal change of disease resistance and CCA1 expression in Arabidopsis. Plant Physiol Biochem 100:150–155. https://doi.org/10.1016/j.plaphy.2016.01.018

Tan DX, Hardeland R, Manchester LC, Korkmaz A, Ma S, Rosales-Corral S, Reiter RJ (2012) Functional roles of melatonin in plants, and perspectives in nutritional and agricultural science. J Exp Bot 63(2):577–597. https://doi.org/10.1093/jxb/err256

van Tassel D, O'Neill S (1993) Melatonin: identification of a potential dark signal in plants. Plant Physiol 102(supplement 1):659

Wang P, Yin L, Liang D, Li C, Ma F, Yue Z (2012) Delayed senescence of apple leaves by exogenous melatonin treatment: toward regulating the ascorbate-glutathione cycle. J Pineal Res 53:11–20

Wang Y, Reiter RJ, Chan Z (2018) Phytomelatonin: a universal abiotic stress regulator. J Exp Bot 69:963–974

Wei J, Li D, Zhang J, Shan C, Rengel Z, Song Z, Chen Q (2018) Phytomelatonin receptor PMTR1-mediated signaling regulates stomatal closure in *Arabidopsis thaliana*. J Pineal Res 65:e12500

Wolf K, Kolar J, Witters E, van Dongen W, van Onckelen H, Machackova I (2001) Daily profile of melatonin levels in *Chenopodium rubrum* L. depends on photoperiod. J Plant Physiol 158: 1491–1493

Xu T, Chen Y, Kang H (2019) Melatonin is a potential target for improving post-harvest preservation of fruits and vegetables. Front Plant Sci 10. https://doi.org/10.3389/fpls.2019.01388

Yu Y, Lv Y, Shi Y, Li T, Chen Y, Zhao D, Zhao Z (2018) The role of phyto-melatonin and related metabolites in response to stress. Molecules 23:1887

Zeng H, Bai Y, Wei Y, Reiter RJ, Shi H (2022a) Phytomelatonin as a central molecule in plant disease resistance. J Exp Bot 73(17):5874–5885. https://doi.org/10.1093/jxb/erac111

Zeng H, Liu M, Wang X, Liu L, Wu H, Chen X, Wang H, Shen Q, Chen G, Wang Y (2022b) Seed-soaking with melatonin for the improvement of seed germination, seedling growth, and the antioxidant defense system under flooding stress. Agronomy 12(8):1918. https://doi.org/10.3390/agronomy12081918

Zhan H, Nie X, Zhang T, Li S, Wang X, Du X, Tong W, Weining S (2019) Melatonin: a small molecule but important for salt stress tolerance in plants. Int J Mol Sci 20:709

Zhang N, Zhao B, Zhang HJ, Weeda S, Yang C, Yang ZC, Ren S, Guo YD (2013) Melatonin promotes water-stress tolerance, lateral root formation, and seed germination in cucumber (*Cucumis sativus* L.). J Pineal Res 54(1):15–23. https://doi.org/10.1111/j.1600-079X.2012.01015

Zhang H, Zhang N, Yang R, Wang L, Sun Q, Li D, Cao Y, Weeda S, Zhao B, Ren S, Guo Y (2014) Melatonin promotes seed germination under high salinity by regulating antioxidant systems, ABA and GA_4 interaction in cucumber (*Cucumis sativus* L.). J Pineal Res 57(3):269–279. https://doi.org/10.1111/jpi.12167

Zhao D, Wang H, Chen S, Yu D, Reiter RJ (2021a) Phytomelatonin: an emerging regulator of plant biotic stress resistance. Trends Plant Sci 26(1):70–82. https://doi.org/10.1016/j.tplants.2020.08.009

Zhao D, Yao Z, Zhang J, Zhang R, Mou Z, Zhang X, Li Z, Feng X, Chen S, Reiter RJ (2021b) Melatonin synthesis genes N-acetylserotonin methyltransferases evolved into caffeic acid O-methyltransferases and both assisted in plant terrestrialization. J Pineal Res 71:e12737

Chapter 2
Melatonin: A Critical Regulator of Plant Growth and Development

Savita Bhardwaj, Dhriti Kapoor, and Neeta Raj Sharma

Abstract Plant growth regulators (PGRs) utilization in the field of agriculture has been widely studied for improving crop quality and yield. Among these PGRs, melatonin (*N*-acetyl-5-methoxy-tryptamine) is an indole moiety, which is mostly present in animals and plants. In animals, the modulation of circadian rhythms and in plants senescence regulation indicate melatonin advantages. It is considered as a versatile moiety because it exerts a positive influence in improving crop quality by improving the morphological parameters, photosynthetic activity, plant metabolism, osmolytes, and the plant's defense system. It can regulate both reactive oxygen and nitrogen species hence regarded as an antioxidant agent. The expression of several stress-specific genes and antioxidant enzyme genes was also upregulated by its application to endure stress resistance in plant genotypes. Melatonin also shows beneficial effects in association with other phytohormones and regulates the gene expression associated with plant hormone metabolism. This book chapter gives an overview of the biosynthesis of melatonin and its role in regulating various aspects of plant growth and development. It also emphasizes the influence of melatonin in the mitigation of various abiotic stresses and its crosstalk with other plant growth regulators in mediating plant growth and development.

Keywords Melatonin · Growth · Antioxidant · Plant growth regulators · Yield

---

S. Bhardwaj
Department of Botany, MCM DAV College, Kangra, Himachal Pradesh, India

Department of Botany, School of Bioengineering and Biosciences, Lovely Professional University, Phagwara, Punjab, India

D. Kapoor (✉)
School of Biological and Environmental Sciences, Shoolini University, Solan, Himachal Pradesh, India

N. R. Sharma
School of Bioengineering and Biosciences, Lovely Professional University, Phagwara, Punjab, India

A. Sharma, G. J. Ahammed (eds.), *Melatonin in Plants: Role in Plant Growth, Development, and Stress Response*, Plant Life and Environment Dynamics,
https://doi.org/10.1007/978-981-99-8051-2_2

2.1 Introduction

Melatonin is a widely distributed moiety whose role has been studied in various organisms such as bacteria, mammals, birds, amphibians, reptiles, fish, and plants. Melatonin shows pleiotropic biological functions, making it a ubiquitous bio-molecule (Hardeland et al. 2011; Nawaz et al. 2016). Several plant genotypes contain melatonin and most of them were shown to have abundant melatonin reserves in their tissues (Nawaz et al. 2016). It also performs multifunctional roles in regulating various developmental aspects such as growth attributes, photosynthetic pigments, gas exchange attributes, senescence, etc. (Altaf et al. 2022; Tiwari et al. 2020; Debnath et al. 2020).

It has been studied in higher plants and other autotrophic species after it was first isolated and explicated in a photosynthetic organism like *Gonyaulax polyedra* (Balzer and Hardeland 1996; Manchester et al. 2000). It also helps plants in overcoming different environmental constraints during their entire life span by stimulating various defense systems in plants thus, recognized as a master plant regulator (Arnao and Hernández-Ruiz 2019; Posmyk and Janas 2009). It functions as a signaling moiety in plants and modulates various biochemical and physiological processes (Sadak et al. 2020). Inhibition of growth and aging, which consequently cause plant deprivation, is caused by various abiotic and biotic constraints. Plants produce melatonin against these stress conditions and exerts a positive role to protect them as that of other PGRs (Martinez et al. 2018; Li et al. 2019).

Plants and animals release free radicals by either photosynthesis or metabolism. In context to this, detoxification of these radicals by melatonin has been shown by previous research dealing with melatonin isolation (Galano and Reiter 2018). The regulation of ROS and RNS by melatonin shows its pleiotropic role as an antioxidant as revealed by earlier studies (Fan et al. 2018; Tan et al. 2003). It regulates plant's antioxidant system to perform ROS and RNS scavenging. It prevents lipid peroxidation in plants by diminishing e^- outflow and by enhancing the ETC efficiency (Reiter et al. 2007; Zhao et al. 2019). Thus melatonin occurrence in seeds, fruits, and reproductive tissues protects them from stress-induced oxidative stress (Wang et al. 2022b).

2.2 Biosynthesis of Melatonin

The pineal gland is the major location for its secretion in animals. However, plants lack such organs, hence melatonin is synthesized differently in plants as compared to animals. Various plant tissues like root, stem, and leaf contain melatonin in them (Fan et al. 2018). Light is the major factor for the regulation of melatonin synthesis in plants and, moreover, mitochondria and chloroplast are the chief locations for its biogenesis (Byeon et al. 2012). These organelles include various types of enzymes that use synthetic processes to produce melatonin. If its synthesis is stopped in

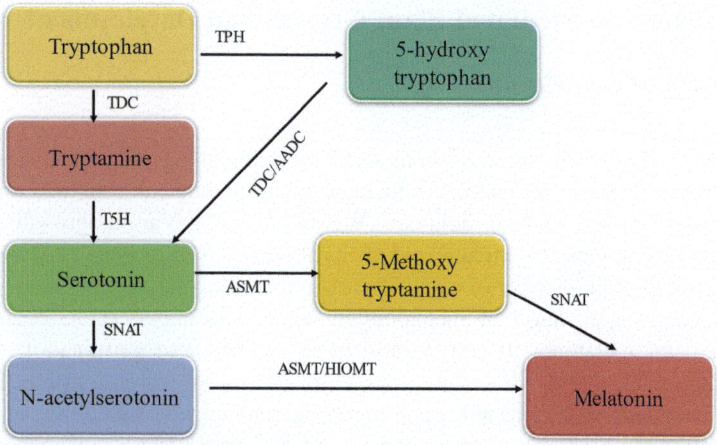

Fig. 2.1 Biosynthesis pathway of melatonin in plants. *TDC* tryptophan decarboxylase, *TPH* tryptophan hydroxylase, *T5H* tryptamine 5-hydroxylase, *SNAT* serotonin *N*-acetyltransferase, *ASMT* N-acetyl serotonin methyltransferase, *HIOMT* hydroxy indole-O-methyltransferase, *AADC* aromatic-L-amino-acid decarboxylase

chloroplasts, then it will begin in mitochondria, indicating that melatonin biosynthesis is a two-way process. Its biogenesis is also stimulated by various environmental constraints and by several factors, for instance, leaf development, fruit maturation, and senescence (Fan et al. 2018). Melatonin 3 hydroxylase (M3H), a specific enzyme is involved in melatonin production (Acharya et al. 2021). The first step in melatonin production is tryptamine synthesis from tryptophan by the tryptophan decarboxylase (TDC) enzyme (Zhou et al. 2020) (Fig. 2.1). Tryptamine to 5-hydroxytryptamine is transformed by tryptamine-5-hydroxylase (T5H) (Kang et al. 2007). Tryptamine is converted into serotonin by T5H which finally gives rise to melatonin.

Another pathway for its synthesis is the formation of 5-hydroxytryptophan by catalyzing tryptophan-5-hydroxylase from tryptophan, as reported in some genotypes like *Hypericum perforatum*. Then serotonin is formed from 5-hydroxytrytophan by TDC/AADC (aromatic-L-amino-acid decarboxylase) which subsequently transformed into *N*-acetyl-serotonin by serotonin *N*-acetyltransferase (SNAT)/aryl alkylamine *N*-acetyltransferase (AANAT). In the end, melatonin is produced from *N* acetyl-serotonin by *N*-acetyl-serotonin methyltransferase (ASMT)/hydroxy indole-*O*-methyltransferase (HIOMT). Formation of 5-methoxytryptamine via HIOMT from serotonin is another way which forms melatonin by SNAT (Fan et al. 2018).

2.3 Melatonin Mediated Plant Growth and Development

2.3.1 *Plant Growth and Biomass*

Melatonin is known to increase final yield by maintaining ion homeostasis and growth traits (Sadak et al. 2020). An increase in morphological traits was observed in melatonin applied soybean cultivars (Wei et al. 2015). Various growth aspects were improved in melatonin treated old plants of *Glycyrrhiza uralensis* (Afreen et al. 2006). Root growth, number of ears and carbon assimilation rate were increased by the exogenous application of melatonin in winter wheat (Ye et al. 2020). The application of melatonin to cotton resulted in an increased germination rate by regulating the GA_3 and ABA ratio (Xiao et al. 2019). An increase in root growth and biomass production was noticed in melatonin applied wheat seedlings under Cd stress (Ni et al. 2018). Root development and content of IAA were improved in *Brassica juncea* by melatonin at minute concentrations (Chen et al. 2009). The rate of seed germination was higher in melatonin treated red cabbage seedlings (Posmyk et al. 2008). Exogenous application of melatonin increased the growth traits and biomass in *Arabidopsis thaliana* genotypes (Pelagio-Flores et al. 2011).

Root development and biomass were increased by melatonin application to rapeseed seedlings by escalating the content of osmolytes and the functioning of antioxidant biomass (Liu et al. 2018). Thus, energy metabolism and C and N metabolism and regulation of ROS regulation are all involved in melatonin mediated plant growth. Seed germination was improved in melatonin treated cucumber seeds by promoting energy production and proteins related to glucose metabolism under NaCl stress (Zhang et al. 2017a). The growth of corn plants was promoted by melatonin by regulating the genes and enzymes related to with C and N metabolism (Erdal 2019). Polyamine oxidase (PAO) dependent H_2O_2 production was stimulated by melatonin in tomatoes which ultimately improved the lateral root development, indicating the H_2O_2 role in growth regulation (Chen et al. 2019). Lateral root formation was induced in alfalfa by melatonin in which H_2O_2 performed as a downstream moiety (Chen et al. 2018). Wheat yield was increased by melatonin application through increasing the functioning of enzymes connected to nitrogen absorption and metabolism (Qiao et al. 2019).

2.3.2 *Photosynthetic Activity*

Melatonin prevents the deprivation of chlorophyll by acting as an antioxidant in plants and protects from ROS/RNS induced senescence. Chlorophyll content was found to be enhanced in melatonin supplemented barley leaves (Arnao and Hernández-Ruiz 2009). Indoleamine of melatonin amended the level of chlorophyll by performing the radical scavenging activity against metal stresses (Tal et al. 2011). Similarly, photosystem II (PS II) efficiency and ascorbic acid measures were also augmented by indoleamine of melatonin in addition to chlorophyll protection against stress circumstances (Wang et al. 2012). The photosynthetic functioning of chlorophyll has also been found to be improved in plants through maintaining redox homeostasis, in addition to aiding in chlorophyll protection (Sharif et al. 2018). Accumulation of melatonin prevents damage to the photosynthetic machinery to overall improve the yield of *Hordeum vulgare*, *Triticum aestivum*, sweet cherry, etc. (Sharma and Zheng 2019). Photosynthesis functioning was noticed to be improved in melatonin supplemented cucumber seedlings by improving the Pn while reducing the CO_2 concentration under heat stress conditions (Xu 2010).

Arabidopsis thaliana transgenics exhibited a maintained redox homeostasis after increasing the synthesis of endogenous melatonin in chloroplasts under salt stress. Furthermore, photosynthetic efficiency and biomass were improved by high melatonin content and low ROS levels to overall improve the NaCl endurance of *A. thaliana* lines (Zheng et al. 2017). In transgenic bermuda grass, melatonin supplementation increased the total chlorophyll content, chlorophyll biosynthesis, PS II efficiency, and expression of photosynthesis-related genes in response to NaCl stress (Shi et al. 2015). Melatonin supplementation increased the gas exchange attributes, i.e., Pn, Gs, and Tr was noticed in maize seedlings in response to limited water availability conditions (Ye et al. 2016).

2.3.3 Leaf Senescence and Fruit Maturation

Various internal and external factors influence leaf senescence among which cytokinin is known to delay leaf senescence while ABA quickens it. Leaf senescence is caused due to the decrease and disruption of chlorophyll and leaf senescence is recognized to be delayed by melatonin. Leaf senescence was noticed to be delayed by melatonin treatment in barley by diminishing the damage to chlorophyll (Arnao and Hernández-Ruiz 2009). During leaf deprivation, levels of chlorophyll and net photosynthetic rate were improved in melatonin applied maize leaves (Ahmad et al. 2020). Exogenously supplemented melatonin increased the growth attributes and induced the cytokinin synthesis in ryegrass by suppressing the expression of senescence-related genes and ABA signal transduction genes in response to high temperature (Zhang et al. 2017b). In creeping bentgrass, drought induced leaf senescence was delayed by melatonin and cytokinin (Ma et al. 2018). Suppression in the activity of ABF-mediated ABA formation was found in melatonin treated genotypes to inhibit leaf aging (Tan et al. 2019). A decrease in the chlorophyll deprivation genes and leaf aging-linked genes was found in melatonin treated tomato genotypes which were initially increased by MeJA (Wang et al. 2019).

Fruit ripening is also regulated by melatonin in plants. Fruit ripening was regulated by melatonin in tomatoes by augmenting the apoptotic inhibitor (API5) proteins and anthocyanin accumulation-related proteins. Moreover, melatonin application also decreased the activity of antioxidant enzyme proteins and senescence-related proteins (Sun et al. 2016). Quality and hardness of peach fruit were promoted by melatonin supplementation by regulating amount of AsA and by delaying the fruit senescence and also by diminishing the fruit's rot rate (Gao et al. 2016). Post-harvest decay of strawberry fruits was decreased by melatonin application through augmenting the content of total phenols and anthocyanins, ATP action, and antioxidant activity (Aghdam and Fard 2017). Contents of ABA, H_2O_2, and ethylene were increased by melatonin application in grape fruits which ultimately resulted in improved fruit ripening, indicating the significance of ethylene and other signaling moieties in controlling fruit ripening (Xu et al. 2018).

2.4 Melatonin Mediated Abiotic Stress Tolerance in Plants

Application of melatonin improves plant tolerance to ROS burst by functioning as a powerful antioxidant to regulate the activities of various antioxidant enzymes and by performing ROS detoxification mechanism (Zhang et al. 2022) (Fig. 2.2). The formation of non-enzymatic antioxidants was also improved by melatonin to alleviate toxic effects of different ecological constraints (Zhang et al. 2022). In maize seedlings, melatonin enhanced the RWC, gas exchange parameters and stimulated the antioxidant defense system to diminish the levels of H_2O_2 and MDA under limited water availability circumstances (Ye et al. 2016). Water deficit endurance was raised in tobacco by melatonin through augmenting the growth attributes, nutrient acquisition, photosynthetic activity, and antioxidant potential via diminishing the levels of H_2O_2 and MDA (Liu et al. 2021). Treatment of maize cultivars with melatonin caused increase in seed germination, growth attributes, biomass, RWC, total phenolic, and osmolyte contents and activities of enzymatic

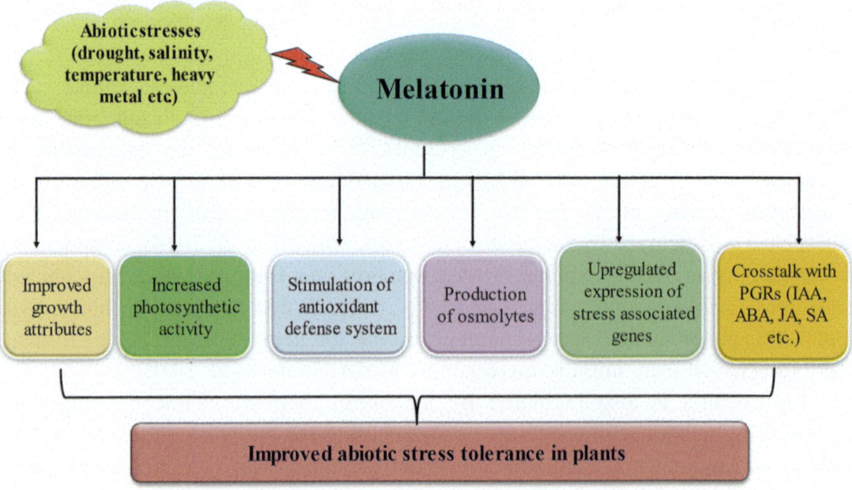

Fig. 2.2 Role of melatonin in improving abiotic stress tolerance in plants

antioxidants. Apart from this, melatonin treatment also diminished the electrolyte leakage by reducing the MDA release under NaCl stress (Jiang et al. 2016).

Similarly, melatonin raised the morphological traits, biomass, photosynthetic pigments, gas exchange attributes, and nutrient homeostasis and stimulated the antioxidant potential in *Phaseolus vulgaris* under salinity environments (Azizi et al. 2022). The application of melatonin markedly mitigated the ROS burst in naked oat seedlings by augmenting the SOD, POD, and CAT enzymes functioning and photosynthetic activity under Pb stress (Wang et al. 2022a). Similarly, melatonin significantly mitigated the As-induced toxicity in rice cultivars (Samanta et al. 2021, 2022). Improved seed germination and seedlings growth was found in melatonin supplemented genotypes by increasing the functioning of α-amylase and protease enzymes in seeds of rice and abridged the levels of MDA, H_2O_2, and $O_2^{\cdot-}$ against Cu stress (Li et al. 2022). The effect of melatonin in different plant species against various stresses has been given in Table 2.1.

2.4.1 Melatonin Crosstalk and Interaction with Other Plant Growth Regulators

Root growth and development were positively regulated by melatonin in rice by stimulating the ROS regulation and the auxin cascade pathway under low oxygenic conditions (Liu et al. 2023). Contents of auxin and their synthesis and transport were augmented by melatonin and various growth and physiological aspects were improved by the interaction of melatonin with auxin. However, reduced melatonin content in plants results in a decline in the auxin concentration in the later growth stages (Ahmad et al. 2023). Combined application of melatonin and GA_3 to drought exposed rapeseed showed an increase in the morphological parameters, proline accumulation, stimulation of antioxidant enzymes to perform ROS scavenging (Khan et al. 2020). Similarly, melatonin and GA_3 in combination increased the morphological traits, and chlorophyll level by escalating the activity of chlorophyll synthesizing enzyme, i.e., δ-aminolevulinic acid dehydratase in tomato under NaCl stress. Furthermore, increase in the osmolytes content and upregulation in the activities of CAT, APX, GR, DHAR, MDHAR, GPOX, and PPO enzymes were also noticed in melatonin and GA_3 applied seedlings under stressed conditions (Siddiqui et al. 2020).

Melatonin and jasmonic acid act synergistically in improving cold stress endurance in tomato by augmenting the Fv/Fm and by diminishing the electrolyte leakage and H_2O_2 content (Ding et al. 2022). Application of melatonin and SA to Zn exposed safflower, improved the fresh and dry weight, photosynthetic pigments, and synthesis of phytochelatins (Eisalou et al. 2021). Similarly, melatonin and SA application both individually as well as in combination significantly alleviated the Cd-induced toxicity in safflower seedlings by improving the plant biomass production, chlorophyll content, decreased amounts of MDA and H_2O_2, and stimulation of Asa-GSH

Table 2.1 Effect of melatonin on different plant species under various abiotic stresses

Stress	Plant species	Melatonin concentration	Results	Reference
Drought stress	*Brassica napus*	0.05 mmol/L	Melatonin increased the growth attributes and CAT, APX, and POD enzyme activities in rapeseed seedlings in response to water deficit conditions	Li et al. (2018)
Drought stress	*Brassica napus*	100 μM	Supplementation of melatonin improved the root growth, stomatal opening, gas exchange attributes, and CAT activity in drought exposed rapeseed genotypes	Dai et al. (2020)
Drought stress	*Glycine max*	100 μM	Soybean seedlings treated with melatonin exhibited increase in growth attributes, plant biomass, photosynthesis efficiency, and antioxidant capacity by reducing lipid peroxidation under stressed conditions	Imran et al. (2021)
Drought stress	*Zea mays*	50 and 100 μM	Melatonin significantly alleviated the drought stress in maize seedlings by improving the growth attributes, plant biomass, photosynthetic activity and stimulated the antioxidant system	Ahmad et al. (2019)
Salinity stress	*Brassica juncea*	1 μM	An increase in the plant growth parameters, gas exchange attributes, RWC, protein, and activities of antioxidative enzymes were found in melatonin supplemented mustard in response to NaCl stress	Park et al. (2021)
Salinity stress	*Brassica napus*	30 μmol/L	Exogenous treatment of melatonin increased the growth parameters, plant biomass, solute accumulation, and POD, CAT, and APX enzymes functioning by reducing the H_2O_2 release in salinity exposed rapeseed seedlings	Liu et al. (2018)
Salinity stress	*Brassica napus*	25, 50, and 100 μM	Melatonin exposed rapeseed showed augmentation in the growth parameters, photosynthetic pigments, gas exchange attributes, relative water content, antioxidants, and compatible solutes levels against salinity stress	Mohamed et al. (2020)
Salinity stress	*Ocimum basilicum*	10 μM	Exogenous application of melatonin significantly mitigated the salinity induced toxic effects in basil by increasing the growth	Bahcesular et al. (2020)

(continued)

Table 2.1 (continued)

Stress	Plant species	Melatonin concentration	Results	Reference
			attributes, content of flavonoids, and phenolic acids	
Salinity stress	*Pistacia vera*	100 μmol/L	An increase in morphological traits, content of photosynthetic pigments, nutrient uptake, osmolytes, and stimulation of antioxidant enzymes was reported in melatonin supplemented pistachio seedlings under NaCl stress	Kamiab (2020)
Salinity stress	*Arachis hypogaea*	50, 100 and 150 μM	Application of melatonin to peanut seedlings diminished the overaccumulation of ROS by activating the antioxidants and also increased the expression of melatonin synthesis associated genes under stressed conditions	ElSayed et al. (2020)
Heat stress	*Triticum aestivum*	100 μM	Melatonin applied wheat exhibited increase in photosynthesis, carbohydrate metabolism, activities of antioxidative enzymes and diminished the ROS burst by diminishing the levels of H_2O_2 and TBARS under stressed conditions	Iqbal et al. (2021)
Heat stress	*Festuca arundinaceous*	20 μM	Melatonin treated tall fescue plants against heat stress showed increase in morphological parameters, photosynthetic pigments, protein content, and functioning of antioxidative enzymes by lowering the MDA and H_2O_2 content	Alam et al. (2022)
Cd and Al stress	*Brassica napus*	50 and 100 μm	Melatonin enhanced the rapeseed endurance to Cd and Al toxicity by escalating the plant growth parameters, biomass, photosynthetic pigments, and activities of antioxidative enzymes	Sami et al. (2020)
Cr stress	*Brassica napus*	10 μM	Morphological traits, photosynthetic pigments, PS II efficiency, osmoprotectants level, and antioxidant enzymes were augmented in melatonin supplemented *B. napus* cultivars against Cr toxicity	Ayyaz et al. (2021)
Cobalt (Co) stress	*Brassica napus*	50 and 100 μM	Exogenous application of melatonin significantly minimized the Co induced toxicity in *B. napus* plants by increasing the plant height, leaf area, gas exchange parameters, osmotic adjustment, and	Ali et al. (2023)

(continued)

Table 2.1 (continued)

Stress	Plant species	Melatonin concentration	Results	Reference
			antioxidant potential. Melatonin treated plants also exhibited reduced lipid peroxidation by inhibiting the ROS overaccumulation	
Cd stress	*Brassica juncea*	50, 70, 75, and 100 M	An increase in growth parameters, photosynthetic activity, biochemical attributes, and antioxidant enzymes was observed in melatonin applied mustard cultivars to significantly reduce Cd toxicity in them	Zargar et al. (2022)

cycle (Amjadi et al. 2021). An increase in plant biomass, photosynthetic pigments, leaf proline amount, phytochelatins levels, PSII efficiency, and augmented nitrogen metabolism was found in melatonin and SA-applied pepper cultivars under As stress (Kaya et al. 2022a).

2.4.2 Melatonin Crosstalk and Interaction with Gasotransmitters

Salinity stress induced the accumulation of melatonin which subsequently improved salinity tolerance in rapeseed seedlings by inducing a signaling cascade to stimulate NR- and NOA1-dependent nitric oxide (NO) amount. Furthermore, *SOS*-mediated ion and redox homeostasis were reestablished and melatonin also triggered NO-dependent *S*-nitrosylation in response to stressed environments (Zhao et al. 2018). Cd exposed *Catharanthus roseus* exhibited an increase in proline content and upregulation of SOD, CAT, APX, and POD enzymes by melatonin and NO pretreatment. Moreover, the application of melatonin and NO decreased lipid peroxidation by declining the H_2O_2 levels under stressed conditions (Nabaei and Amooaghaie 2019). Co-application of melatonin and NO to Cd stressed *C. roseus* plants significantly increased the shoot biomass, pigments, and mineral homeostasis and diminished lipid peroxidation. In addition to this, melatonin and NO treatment augmented the plant biomass and Cd accumulation and transport from root to upper plant parts to improve the phytoremediation efficiency of Cd (Nabaei and Amooaghaie 2020).

Combined treatment of melatonin and NO augmented the growth attributes, plant biomass, and photosynthetic activity of soybean under Pb and Cd stress. Endogenous abscisic acid and jasmonic acid and Pb and Cd mobilization were diminished in melatonin and NO applied soybean plants (Imran et al. 2022). Melatonin and NO synergistically lessened the NaCl stress in sunflower seedlings by modulating seedling growth and oxidative homeostasis (Arora and Bhatla 2017). Application of melatonin to high temperature exposed wheat escalated the photosynthetic activity, and carbohydrate activity and decreased the ROS burst by escalating the functioning of antioxidants in association with H_2S (Iqbal et al. 2021). Kaya et al. (2020) indicated that melatonin application suggestively amplified plant growth, and stimulated the defense system to mitigate the salinity stress and iron deficiency in *Capsicum annuum* by involving downstream signal crosstalk with NO and H_2S. Melatonin and H_2S applied *C. annuum* plants under As stress exhibited increased plant height and leaf size whereas ROS burst was lessened by upregulating the antioxidant and glyoxalase systems (Kaya et al. 2022b).

2.5 Conclusion

Melatonin is recognized as a multifunctional growth moiety in plants that has the potential for regulating various morphological, biochemical, and physiological aspects. Various toxic effects induced by abiotic stresses in different plants have been mitigated by the use of melatonin by increasing various developmental aspects. Plant metabolism and the intricate mechanism of crop functions were significantly modulated by melatonin but the exact mechanism of the melatonin action under stressed conditions is still poorly understood. In addition to this, the interaction of melatonin with other PGRs and gasotransmitters also exhibits a positive role in increasing crop productivity and yield in response to stressed conditions. Melatonin also upregulates the expression of various defense-related genes at the molecular level. It can be concluded that melatonin may prove to be a significant molecule to influence plant growth and may prove significant in elevating crop productivity and the nutraceutical value to increase food security issues around the world.

References

Acharya D, Satapathy S, Somu P, Parida UK, Mishra G (2021) Apoptotic effect and anticancer activity of biosynthesized silver nanoparticles from marine algae *Chaetomorpha linum* extract against human colon cancer cell HCT-116. Biol Trace Elem Res 199:1812–1822

Afreen F, Zobayed SM, Kozai T (2006) Melatonin in *Glycyrrhiza uralensis*: response of plant roots to spectral quality of light and UV-B radiation. J Pineal Res 41(2):108–115

Aghdam MS, Fard JR (2017) Melatonin treatment attenuates postharvest decay and maintains nutritional quality of strawberry fruits (Fragaria× anannasa cv. Selva) by enhancing GABA shunt activity. Food Chem 221:1650–1657

Ahmad S, Kamran M, Ding R, Meng X, Wang H, Ahmad I, Fahad S, Han Q (2019) Exogenous melatonin confers drought stress by promoting plant growth, photosynthetic capacity and antioxidant defense system of maize seedlings. PeerJ 7:e7793

Ahmad S, Su W, Kamran M, Ahmad I, Meng X, Wu X, Javed T, Han Q (2020) Foliar application of melatonin delay leaf senescence in maize by improving the antioxidant defense system and enhancing photosynthetic capacity under semi-arid regions. Protoplasma 257:1079–1092

Ahmad I, Song X, Hussein Ibrahim ME, Jamal Y, Younas MU, Zhu G, Zhou G, Adam Ali AY (2023) The role of melatonin in plant growth and metabolism, and its interplay with nitric oxide and auxin in plants under different types of abiotic stress. Front Plant Sci 14:1108507

Alam MN, Yang L, Yi X, Wang Q, Robin AHK (2022) Role of melatonin in inducing the physiological and biochemical processes associated with heat stress tolerance in tall fescue (*Festuca arundinaceous*). J Plant Growth Regul 41(7):2759–2768

Ali S, Gill RA, Ulhassan Z, Zhang N, Hussain S, Zhang K, Huang Q, Sagir M, Tahir MB, Gill MB (2023) Exogenously applied melatonin enhanced the tolerance of *Brassica napus* against cobalt toxicity by modulating antioxidant defense, osmotic adjustment, and expression of stress response genes. Ecotoxicol Environ Saf 252:114624

Altaf MA, Shahid R, Altaf MM, Kumar R, Naz S, Kumar A, Alam P, Tiwari RK, Lal MK, Ahmad P (2022) Melatonin: first-line soldier in tomato under abiotic stress current and future perspective. Plant Physiol Biochem 185:188–197

Amjadi Z, Namdjoyan S, Abolhasani Soorki A (2021) Exogenous melatonin and salicylic acid alleviates cadmium toxicity in safflower (*Carthamus tinctorius* L.) seedlings. Ecotoxicology 30: 387–401

Arnao M, Hernández-Ruiz J (2009) Protective effect of melatonin against chlorophyll degradation during the senescence of barley leaves. J Pineal Res 46(1):58–63

Arnao MB, Hernández-Ruiz J (2019) Melatonin: a new plant hormone and/or a plant master regulator? Trends Plant Sci 24(1):38–48

Arora D, Bhatla SC (2017) Melatonin and nitric oxide regulate sunflower seedling growth under salt stress accompanying differential expression of cu/Zn SOD and Mn SOD. Free Radic Biol Med 106:315–328

Ayyaz A, Farooq MA, Dawood M, Majid A, Javed M, Athar HR, Bano H, Zafar ZU (2021) Exogenous melatonin regulates chromium stress-induced feedback inhibition of photosynthesis and antioxidative protection in *Brassica napus* cultivars. Plant Cell Rep 40(11):2063–2080

Azizi F, Amiri H, Ismaili A (2022) Melatonin improves salinity stress tolerance of *Phaseolus vulgaris* L. cv. Pak by changing antioxidant enzymes and photosynthetic parameters. Acta Physiol Plant 44(4):40

Bahcesular B, Yildirim ED, Karaçocuk M, Kulak M, Karaman S (2020) Seed priming with melatonin effects on growth, essential oil compounds and antioxidant activity of basil (*Ocimum basilicum* L.) under salinity stress. Ind Crop Prod 146:112165

Balzer I, Hardeland R (1996) Melatonin in algae and higher plants-possible new roles as a phytohormone and antioxidant. Bot Acta 109(3):180–183

Byeon Y, Park S, Kim YS, Park DH, Lee S, Back K (2012) Light-regulated melatonin biosynthesis in rice during the senescence process in detached leaves. J Pineal Res 53(1):107–111

Chen Q, Qi W-b, Reiter RJ, Wei W, Wang B-m (2009) Exogenously applied melatonin stimulates root growth and raises endogenous indoleacetic acid in roots of etiolated seedlings of *Brassica juncea*. J Plant Physiol 166(3):324–328

Chen Z, Gu Q, Yu X, Huang L, Xu S, Wang R, Shen W, Shen W (2018) Hydrogen peroxide acts downstream of melatonin to induce lateral root formation. Ann Bot 121(6):1127–1136

Chen J, Li H, Yang K, Wang Y, Yang L, Hu L, Liu R, Shi Z (2019) Melatonin facilitates lateral root development by coordinating PAO-derived hydrogen peroxide and Rboh-derived superoxide radical. Free Radic Biol Med 143:534–544

Dai L, Li J, Harmens H, Zheng X, Zhang C (2020) Melatonin enhances drought resistance by regulating leaf stomatal behaviour, root growth and catalase activity in two contrasting rapeseed (*Brassica napus* L.) genotypes. Plant Physiol Biochem 149:86–95

Debnath B, Li M, Liu S, Pan T, Ma C, Qiu D (2020) Melatonin-mediate acid rain stress tolerance mechanism through alteration of transcriptional factors and secondary metabolites gene expression in tomato. Ecotoxicol Environ Saf 200:110720

Ding F, Ren L, Xie F, Wang M, Zhang S (2022) Jasmonate and melatonin act synergistically to potentiate cold tolerance in tomato plants. Front Plant Sci 12:3193

Eisalou AV, Namdjoyan S, Soorki AA (2021) The combined effect of melatonin and salicylic acid improved the tolerance of safflower seedlings to zinc toxicity. Acta Physiol Plant 43:1–14

ElSayed AI, Boulila M, Rafudeen MS, Mohamed AH, Sengupta S, Rady M, Omar AA (2020) Melatonin regulatory mechanisms and phylogenetic analyses of melatonin biosynthesis related genes extracted from peanut under salinity stress. Plan Theory 9(7):854

Erdal S (2019) Melatonin promotes plant growth by maintaining integration and coordination between carbon and nitrogen metabolisms. Plant Cell Rep 38:1001–1012

Fan J, Xie Y, Zhang Z, Chen L (2018) Melatonin: a multifunctional factor in plants. Int J Mol Sci 19(5):1528

Galano A, Reiter RJ (2018) Melatonin and its metabolites vs oxidative stress: from individual actions to collective protection. J Pineal Res 65(1):e12514

Gao H, Zhang ZK, Chai HK, Cheng N, Yang Y, Wang DN, Yang T, Cao W (2016) Melatonin treatment delays postharvest senescence and regulates reactive oxygen species metabolism in peach fruit. Postharv Biol Technol 118:103–110

Hardeland R, Cardinali DP, Srinivasan V, Spence DW, Brown GM, Pandi-Perumal SR (2011) Melatonin—a pleiotropic, orchestrating regulator molecule. Prog Neurobiol 93(3):350–384

Imran M, Shazad R, Bilal S, Imran QM, Khan M, Kang S-M, Khan AL, Yun B-W, Lee I-J (2021) Exogenous melatonin mediates the regulation of endogenous nitric oxide in *Glycine max* L. to reduce effects of drought stress. Environ Exp Bot 188:104511

Imran M, Khan AL, Mun B-G, Bilal S, Shaffique S, Kwon E-H, Kang S-M, Yun B-W, Lee I-J (2022) Melatonin and nitric oxide: dual players inhibiting hazardous metal toxicity in soybean plants via molecular and antioxidant signaling cascades. Chemosphere 308:136575

Iqbal N, Fatma M, Gautam H, Umar S, Sofo A, D'ippolito I, Khan NA (2021) The crosstalk of melatonin and hydrogen sulfide determines photosynthetic performance by regulation of carbohydrate metabolism in wheat under heat stress. Plan Theory 10(9):1778

Jiang X, Li H, Song X (2016) Seed priming with melatonin effects on seed germination and seedling growth in maize under salinity stress. Pak J Bot 48(4):1345–1352

Kamiab F (2020) Exogenous melatonin mitigates the salinity damages and improves the growth of pistachio under salinity stress. J Plant Nutr 43(10):1468–1484

Kang S, Kang K, Lee K, Back K (2007) Characterization of tryptamine 5-hydroxylase and serotonin synthesis in rice plants. Plant Cell Rep 26:2009–2015

Kaya C, Higgs D, Ashraf M, Alyemeni MN, Ahmad P (2020) Integrative roles of nitric oxide and hydrogen sulfide in melatonin-induced tolerance of pepper (*Capsicum annuum* L.) plants to iron deficiency and salt stress alone or in combination. Physiol Plant 168(2):256–277

Kaya C, Sarioglu A, Ashraf M, Alyemeni MN, Ahmad P (2022a) The combined supplementation of melatonin and salicylic acid effectively detoxifies arsenic toxicity by modulating phytochelatins and nitrogen metabolism in pepper plants. Environ Pollut 297:118727. https://doi.org/10.1016/j.envpol.2021.118727

Kaya C, Ugurlar F, Ashraf M, Alyemeni MN, Bajguz A, Ahmad P (2022b) The involvement of hydrogen sulphide in melatonin-induced tolerance to arsenic toxicity in pepper (*Capsicum annuum* L.) plants by regulating sequestration and subcellular distribution of arsenic, and antioxidant defense system. Chemosphere 309(Pt 1):136678. https://doi.org/10.1016/j.chemosphere.2022.136678

Khan MN, Khan Z, Luo T, Liu J, Rizwan M, Zhang J, Xu Z, Wu H, Hu L (2020) Seed priming with gibberellic acid and melatonin in rapeseed: consequences for improving yield and seed quality under drought and non-stress conditions. Ind Crop Prod 156:112850

Li J, Zeng L, Cheng Y, Lu G, Fu G, Ma H, Liu Q, Zhang X, Zou X, Li C (2018) Exogenous melatonin alleviates damage from drought stress in *Brassica napus* L.(rapeseed) seedlings. Acta Physiol Plant 40:1–11

Li J, Liu J, Zhu T, Zhao C, Li L, Chen M (2019) The role of melatonin in salt stress responses. Int J Mol Sci 20(7):1735

Li R, Wu L, Shao Y, Hu Q, Zhang H (2022) Melatonin alleviates copper stress to promote rice seed germination and seedling growth via crosstalk among various defensive response pathways. Plant Physiol Biochem 179:65–77

Liu Z, Cai J-S, Li J-J, Lu G-Y, Li C-S, Fu G-P, Zhang X-K, Liu Q-Y, Zou X-L, Cheng Y (2018) Exogenous application of a low concentration of melatonin enhances salt tolerance in rapeseed (*Brassica napus* L.) seedlings. J Integr Agric 17(2):328–335

Liu L, Li D, Ma Y, Shen H, Zhao S, Wang Y (2021) Combined application of arbuscular mycorrhizal fungi and exogenous melatonin alleviates drought stress and improves plant growth in tobacco seedlings. J Plant Growth Regul 40:1074–1087

Liu J, Wang J, Zhang T, Li M, Yan H, Liu Q, Wei Y, Ji X, Zhao Q (2023) Exogenous melatonin positively regulates rice root growth through promoting the antioxidant system and mediating the auxin signaling under root-zone hypoxia stress. Agronomy 13(2):386

Ma X, Zhang J, Burgess P, Rossi S, Huang B (2018) Interactive effects of melatonin and cytokinin on alleviating drought-induced leaf senescence in creeping bentgrass (*Agrostis stolonifera*). Environ Exp Bot 145:1–11

Manchester LC, Tan D-X, Reiter RJ, Park W, Monis K, Qi W (2000) High levels of melatonin in the seeds of edible plants: possible function in germ tissue protection. Life Sci 67(25):3023–3029

Martinez V, Nieves-Cordones M, Lopez-Delacalle M, Rodenas R, Mestre TC, Garcia-Sanchez F, Rubio F, Nortes PA, Mittler R, Rivero RM (2018) Tolerance to stress combination in tomato plants: new insights in the protective role of melatonin. Molecules 23(3):535

Mohamed IA, Shalby N, El-Badri MA, Saleem MH, Khan MN, Nawaz M, Qin M, Agami RA, Kuai J, Wang B (2020) Stomata and xylem vessels traits improved by melatonin application contribute to enhancing salt tolerance and fatty acid composition of *Brassica napus* L. plants. Agronomy 10(8):1186

Nabaei M, Amooaghaie R (2019) Nitric oxide is involved in the regulation of melatonin-induced antioxidant responses in *Catharanthus roseus* roots under cadmium stress. Botany 97(12): 681–690

Nabaei M, Amooaghaie R (2020) Melatonin and nitric oxide enhance cadmium tolerance and phytoremediation efficiency in *Catharanthus roseus* (L.) G. Don. Environ Sci Pollut Res 27: 6981–6994

Nawaz MA, Huang Y, Bie Z, Ahmed W, Reiter RJ, Niu M, Hameed S (2016) Melatonin: current status and future perspectives in plant science. Front Plant Sci 6:1230

Ni J, Wang Q, Shah FA, Liu W, Wang D, Huang S, Fu S, Wu L (2018) Exogenous melatonin confers cadmium tolerance by counterbalancing the hydrogen peroxide homeostasis in wheat seedlings. Molecules 23(4):799

Park H-S, Kazerooni EA, Kang S-M, Al-Sadi AM, Lee I-J (2021) Melatonin enhances the tolerance and recovery mechanisms in *Brassica juncea* (L.) Czern. under saline conditions. Front Plant Sci 12:593717

Pelagio-Flores R, Ortíz-Castro R, Méndez-Bravo A, Macías-Rodríguez L, López-Bucio J (2011) Serotonin, a tryptophan-derived signal conserved in plants and animals, regulates root system architecture probably acting as a natural auxin inhibitor in *Arabidopsis thaliana*. Plant Cell Physiol 52(3):490–508

Posmyk MM, Janas KM (2009) Melatonin in plants. Acta Physiol Plant 31:1–11

Posmyk MM, Kuran H, Marciniak K, Janas KM (2008) Presowing seed treatment with melatonin protects red cabbage seedlings against toxic copper ion concentrations. J Pineal Res 45(1): 24–31

Qiao Y, Yin L, Wang B, Ke Q, Deng X, Wang S (2019) Melatonin promotes plant growth by increasing nitrogen uptake and assimilation under nitrogen deficient condition in winter wheat. Plant Physiol Biochem 139:342–349

Reiter RJ, Tan D-x, Manchester LC, Simopoulos AP, Maldonado MD, Flores LJ, Terron MP (2007) Melatonin in edible plants (phytomelatonin): identification, concentrations, bioavailability and proposed functions. World Rev Nutr Diet 97:211–230

Sadak MS, Abdalla AM, Abd Elhamid EM, Ezzo M (2020) Role of melatonin in improving growth, yield quantity and quality of *Moringa oleifera* L. plant under drought stress. Bull Natl Res Centre 44(1):1–13

Samanta S, Banerjee A, Roychoudhury A (2021) Exogenous melatonin regulates endogenous phytohormone homeostasis and thiol-mediated detoxification in two indica rice cultivars under arsenic stress. Plant Cell Rep 40(8):1585–1602

Samanta S, Banerjee A, Roychoudhury A (2022) Arsenic toxicity is counteracted by exogenous application of melatonin to different extents in arsenic-susceptible and arsenic-tolerant rice cultivars. J Plant Growth Regul 41(6):2210–2231

Sami A, Shah F, Abdullah M, Zhou X, Yan Y, Zhu Z, Zhou K (2020) Melatonin mitigates cadmium and aluminium toxicity through modulation of antioxidant potential in *Brassica napus* L. Plant Biol 22(4):679–690

Sharif R, Xie C, Zhang H, Arnao MB, Ali M, Ali Q, Muhammad I, Shalmani A, Nawaz MA, Chen P (2018) Melatonin and its effects on plant systems. Molecules 23(9):2352

Sharma A, Zheng B (2019) Melatonin mediated regulation of drought stress: physiological and molecular aspects. Plan Theory 8(7):190

Shi H, Jiang C, Ye T, Tan D-X, Reiter RJ, Zhang H, Liu R, Chan Z (2015) Comparative physiological, metabolomic, and transcriptomic analyses reveal mechanisms of improved abiotic stress resistance in bermudagrass [*Cynodon dactylon* (L). Pers.] by exogenous melatonin. J Exp Bot 66(3):681–694

Siddiqui MH, Alamri S, Alsubaie QD, Ali HM (2020) Melatonin and gibberellic acid promote growth and chlorophyll biosynthesis by regulating antioxidant and methylglyoxal detoxification system in tomato seedlings under salinity. J Plant Growth Regul 39:1488–1502

Sun Q, Zhang N, Wang J, Cao Y, Li X, Zhang H, Zhang L, Tan DX, Guo YD (2016) A label-free differential proteomics analysis reveals the effect of melatonin on promoting fruit ripening and anthocyanin accumulation upon postharvest in tomato. J Pineal Res 61(2):138–153

Tal O, Haim A, Harel O, Gerchman Y (2011) Melatonin as an antioxidant and its semi-lunar rhythm in green macroalga Ulva sp. J Exp Bot 62(6):1903–1910

Tan DX, Manchester LC, Hardeland R, Lopez-Burillo S, Mayo JC, Sainz RM, Reiter RJ (2003) Melatonin: a hormone, a tissue factor, an autocoid, a paracoid, and an antioxidant vitamin. J Pineal Res 34(1):75–78

Tan XL, Zq F, Kuang Jf L, Wj RRJ, Lakshmanan P, Xg S, Zhou J, Jy C, Shan W (2019) Melatonin delays leaf senescence of Chinese flowering cabbage by suppressing ABFs-mediated abscisic acid biosynthesis and chlorophyll degradation. J Pineal Res 67(1):e12570

Tiwari RK, Lal MK, Naga KC, Kumar R, Chourasia KN, Subhash S, Kumar D, Sharma S (2020) Emerging roles of melatonin in mitigating abiotic and biotic stresses of horticultural crops. Sci Hortic 272:109592

Wang P, Yin L, Liang D, Li C, Ma F, Yue Z (2012) Delayed senescence of apple leaves by exogenous melatonin treatment: toward regulating the ascorbate–glutathione cycle. J Pineal Res 53(1):11–20

Wang M, Zhang T, Ding F (2019) Exogenous melatonin delays methyl jasmonate-triggered senescence in tomato leaves. Agronomy 9(12):795

Wang K, He J, Zhao N, Zhao Y, Qi F, Fan F, Wang Y (2022a) Effects of melatonin on growth and antioxidant capacity of naked oat (*Avena nuda* L) seedlings under lead stress. PeerJ 10:e13978

Wang K, Xing Q, Ahammed GJ, Zhou J (2022b) Functions and prospects of melatonin in plant growth, yield, and quality. J Exp Bot 73(17):5928–5946

Wei W, Li Q-T, Chu Y-N, Reiter RJ, Yu X-M, Zhu D-H, Zhang W-K, Ma B, Lin Q, Zhang J-S (2015) Melatonin enhances plant growth and abiotic stress tolerance in soybean plants. J Exp Bot 66(3):695–707

Xiao S, Liu L, Wang H, Li D, Bai Z, Zhang Y, Sun H, Zhang K, Li C (2019) Exogenous melatonin accelerates seed germination in cotton (*Gossypium hirsutum* L.). PloS One 14(6):e0216575

Xu X (2010) Effect of exogenous melatonin on the physiological response of cucumber seedlings under high temperature stress. Master's Thesis. Northwest Agriculture and Forestry University, Xianyang, China

Xu L, Yue Q, Xiang G, Fe B, Yao Y (2018) Melatonin promotes ripening of grape berry via increasing the levels of ABA, H2O2, and particularly ethylene. Hortic Res 5:41

Ye J, Wang S, Deng X, Yin L, Xiong B, Wang X (2016) Melatonin increased maize (*Zea mays* L.) seedling drought tolerance by alleviating drought-induced photosynthetic inhibition and oxidative damage. Acta Physiol Plant 38(2):48

Ye J, Yang W, Li Y, Wang S, Yin L, Deng X (2020) Seed pre-soaking with melatonin improves wheat yield by delaying leaf senescence and promoting root development. Agronomy 10(1):84

Zargar T, Mir A, Alam P, Hayat S (2022) Melatonin alleviates cadmium-induced toxicity by modulating antioxidant defence mechanisms, growth and photosynthesis in *Brassica juncea*. Russ J Plant Physiol 69(6):1–10

Zhang N, Zhang H-J, Sun Q-Q, Cao Y-Y, Li X, Zhao B, Wu P, Guo Y-D (2017a) Proteomic analysis reveals a role of melatonin in promoting cucumber seed germination under high salinity by regulating energy production. Sci Rep 7(1):1–15

Zhang J, Shi Y, Zhang X, Du H, Xu B, Huang B (2017b) Melatonin suppression of heat-induced leaf senescence involves changes in abscisic acid and cytokinin biosynthesis and signaling pathways in perennial ryegrass (*Lolium perenne* L.). Environ Exp Bot 138:36–45

Zhang T, Wang J, Sun Y, Zhang L, Zheng S (2022) Versatile roles of melatonin in growth and stress tolerance in plants. J Plant Growth Regul 41:1–17

Zhao G, Zhao Y, Yu X, Kiprotich F, Han H, Guan R, Wang R, Shen W (2018) Nitric oxide is required for melatonin-enhanced tolerance against salinity stress in rapeseed (*Brassica napus* L.) seedlings. Int J Mol Sci 19(7):1912

Zhao D, Yu Y, Shen Y, Liu Q, Zhao Z, Sharma R, Reiter RJ (2019) Melatonin synthesis and function: evolutionary history in animals and plants. Front Endocrinol 10:249

Zheng X, Tan DX, Allan AC, Zuo B, Zhao Y, Reiter RJ, Wang L, Wang Z, Guo Y, Zhou J (2017) Chloroplastic biosynthesis of melatonin and its involvement in protection of plants from salt stress. Sci Rep 7(1):1–12

Zhou Y, Liao L, Liu X, Liu B, Chen X, Guo Y, Huang C, Zhao Y, Zeng Z (2020) Crystal structure of Oryza sativa TDC reveals the substrate specificity for TDC-mediated melatonin biosynthesis. J Adv Res 24:501–511

Chapter 3
Regulation of Photosynthesis by Melatonin Under Optimal and Suboptimal Conditions

Anjuman Hussain, Mohammad Faizan, and S. Maqbool Ahmed

Abstract Melatonin is a well-known pleiotropic signalling molecule that controls metabolic functions, photosynthesis, and plant growth. Melatonin's ability to increase plant stress resilience has been studied extensively, but its impact on photosynthesis has received less attention. Here, we have provided a summary of melatonin's function in plants and how it affects photosynthesis in both normal and stressful environments. Melatonin controls the transcription of associated genes and hormone cues to control the synthesis and breakdown of chlorophyll. Through better photosystem gene transcription, activation of the antioxidant system and promotion of the xanthophyll cycle, it safeguards photosynthesis proteins and upholds the photosynthetic process. Through CAND2/PMTR1, melatonin may control plant stomatal mobility. Lastly, it regulates the metabolism of sugar, the gluconeogenesis pathway, and the breakdown and transport of transient starch to control the photosynthetic carbon cycle.

Keywords Melatonin · Photosynthesis · Drought stress · Physiological growth

3.1 Introduction

The mammalian pineal secretory molecule known as melatonin was first identified in the bovine pineal gland (Reiter 1991). The pineal gland is an organ found in animal bodies that produces melatonin to regulate how the body responds to changing photoperiods and also functions as an antioxidant to safeguard neurons (Tan et al. 2018). More than 20 dicotyledonous and monocotyledonous plant families have been found to contain melatonin (Nawaz et al. 2016]. After Lerner and Case (1959)

A. Hussain
Plant Physiology Lab, Department of Botany, Aligarh Muslim University, Aligarh, India

M. Faizan (✉) · S. M. Ahmed
Botany Section, School of Sciences, Maulana Azad National Urdu University, Hyderabad, Telangana, India
e-mail: mohammadfaizan@manuu.edu.in

A. Sharma, G. J. Ahammed (eds.), *Melatonin in Plants: Role in Plant Growth, Development, and Stress Response*, Plant Life and Environment Dynamics,
https://doi.org/10.1007/978-981-99-8051-2_3

claimed that an indole molecule causes skin lighting in the frog, the name melatonin was assigned to these biomolecules. The secretion of melatonin achieves its peak during the night, making it the peak signalling molecule of darkness, and it is also the most significant factor for regulating the circadian cycle in various vertebrates (Reiter 1991). Melatonin is also a significant antioxidant that can be consumed in food and is also produced by the body, though this production progressively declines with ageing (Reiter 1991; Hardeland 2013). Melatonin's function in plants has been thoroughly researched (Arnao and Hernández-Ruiz 2013; Byeon and Back 2014). About 34,000 research papers on melatonin have been published since its discovery in 1965, according to the Scopus database. This emphasizes the molecule's significance by showing how widely it has been researched. It draws scientists and young researchers from the broad area of plant sciences due to its significant effects on plant systems (Reiter et al. 2015; Arnao and Hernández-Ruiz 2015). It is acknowledged as a crucial plant metabolite and is regarded as a central indolamine neurotransmitter, heavily engaged in a variety of biological processes. Melatonin has also been linked to improvements in seed germination, fruit ripening, photosynthesis, biomass production, circadian rhythm, redox network, membrane integrity, root development, leaf senescence, osmoregulation, and resistance to abiotic stress (salt, drought, cold, heat, oxidative, and heavy metals) (Qian et al. 2015; Lee et al. 2015). Additionally, it has been claimed to help plants defend themselves from biotic stressors (Arnao and Hernández-Ruiz 2015). Additionally, according to Zhang et al. (2016), melatonin stimulates gene expression that aids the plant in coping with biotic and abiotic stresses. Therefore, using melatonin as a bio-stimulator for sustainable crop production that does not harm the ecosystem could be very important.

3.2 Role of Melatonin in Plants

Plant tissues have shown to constantly produce melatonin (Wang et al. 2014). There have been reports of pleiotropic functions that range from promoting plant germination to delaying plant senescence (Chan and Shi 2015; Wei et al. 2015). Numerous phytohormones, particularly auxin, played a role in the growth and evolution of plants. Melatonin and IAA share the same precursor, tryptophan, making them both types of indolamine. As such, melatonin should be involved in the control of plant growth and development. In lupin hypocotyls as well as in monocot species such as canary grass, wheat, barley, and oat (Arnao and Hernández-Ruiz 2007) and dicot species like arabidopsis (Chen et al. 2017), melatonin is thought to be a growth-promoting molecule, similar to auxin. As such, it is an auxinic hormone in plants. The concentration of melatonin in *Chenopodium rubrum* was found to fluctuate on a regular basis throughout the 12-h light/12-h day cycle, indicating that melatonin regulates circadian cycles in plants (Kolář et al. 1997). *Chenopodium rubrum* L. flowering after the application of 100 and 500 μM melatonin, seedlings exposed to a single inductive 12-h period of darkness greatly decreased, but there was no discernible change in photoperiodic time. This finding suggested that exogenous melatonin might have an impact on flower growth during the photoperiod's early stages (Kolář et al. 2003). After receiving melatonin treatment, soybean plants' leaf size, plant height, pod count, and seed number all substantially increased, indicating that exogenous melatonin may enhance the growth and seed production of soybean plants (Wei et al. 2015). Melatonin treatment inhibited the degradation of chlorophyll in barley leaves, demonstrating its protective function in the senescence of plants (Arnao and Hernández-Ruiz 2009). The decrease of chlorophyll content and photosystem efficiency (Fv/Fm) in detached apple (*Malus domestica*) leaves were postponed after melatonin treatment, indicating that the dark-induced senescence process was postponed by 10 mM melatonin solution (Wang et al. 2012). It was demonstrated that ascorbate peroxidase (APX) activity was increased and H_2O_2 accumulation was suppressed. When compared to the control, melatonin increased ascorbic acid (AsA) and glutathione (GSH) levels while decreasing dehydroascorbate (DHA) and oxidized glutathione (GSSG), indicating that it controlled plant ageing through the ascorbate-glutathione cycle (Wang et al. 2012). According to Zhang et al. findings, melatonin promoted the development of lateral roots in cucumber plants (*Cucumis sativus*) (Zhang et al. 2014). According to Hernández-Ruiz and Arnao (2008), melatonin may play a role in etiolated cotyledons of lupin (*Lupinus albus* L.) by stimulating their growth, which is similar to how IAA works.

Additionally, melatonin's impacts on plants varied depending on its concentration. Low melatonin concentrations (10–20 μM) had no discernible impact on root length in arabidopsis plants. On the other hand, high melatonin content (200–400 μM) greatly inhibited fresh weight, and the moderate condition (40 μM) was the best for promoting plant growth and development (Bajwa et al. 2014). Due to the degradation, postharvest produce's quality and shelf life are affected. To keep the quality and shelf life of postharvest fruits and vegetables, numerous treatments have been put in place (Sharif et al. 2018a; Cao et al. 2018). The primary disadvantage of cold storage is that produce is typically kept in a cold environment, which increases the production of ROS and causes oxidative stress (Sun et al. 2016). However, melatonin therapy reduces ROS activity and boosts the synthesis of antioxidant enzymes (Cao et al. 2018). In other situations, the application of exogenous melatonin sparked the endogenous melatonin biosynthetic activity through the antagonistic crosstalk with calcium, stopping the product from deteriorating postharvest (Hu et al. 2018). Furthermore, because postharvest quality of horticultural produce can only be maintained after harvesting, it is largely dependent on preharvest factors (Arah et al. 2015). Accordingly, the melatonin-fertigated tomato seeds had not only increased their yield but also maintained the quality of their postharvest by displaying a rise in vitamin C, lycopene, and calcium contents. Additionally, the treated plants showed higher levels of soluble minerals and P content than the control plants (Liu et al. 2016). In a different research, exogenous melatonin treatment of grape clusters on the plant affected polyphenol metabolism, carbohydrate biosynthesis, and most significantly, ethylene signalling in grape berries. Better antioxidant activity was obtained as a consequence of the restricted ethylene production (Hu et al. 2018), which is crucial for preserving postharvest quality. In addition, melatonin controls salicylic acid, jasmonic acid, nitric oxide, and ethylene, which together produce the resilience against diseases in a very well-known manner (Arnao and Hernández-Ruiz 2018). The ethylene biosynthesis pathway was also affected by exogenously applied melatonin, which improved tomato postharvest quality in terms of aroma, colour, sugar, and total postharvest quality (Sun et al. 2015). Melatonin, however, might be taken into account as a possible compound to lower the proportion of postharvest losses and lengthen the shelf life of postharvest produce. In a recent study, it was discovered that suppressing the fruit shelf life regulator (SIFR) gene reduced tomato postharvest ripeness and increased fruit shelf life by reducing ethylene production (Zhang et al. 2018). It will be intriguing to see how exogenous melatonin impacts postharvest maturity by controlling the level of SIFR gene expression.

Melatonin has been shown in studies to have a protective impact on a plant's reproductive tissue, and its concentration varies with the development of the fruit and flower tissue. Tryptophan, a forerunner to the synthesis of melatonin, is accumulated in rice flowers during the production stage, and the amount of melatonin in flower spikes is six times greater than that in the associated flag leaves (Park et al. 2013). Melatonin is a hormone that the mother plant of wheat secretes during the grain-filling stage, and it has been shown to boost antioxidant enzyme activity in the progeny and enhance seedling cold tolerance. Early research revealed that the melatonin content in datura flower buds was highest during the early stages of development and progressively decreased as the flower buds matured. The content maintained its high degree of stability (Murch et al. 2009). Before breeding, the melatonin content in grape skin rose to 45% of fully developed purple (Murch et al. 2010). On the other hand, melatonin levels in the pulp, seeds, and skin are comparatively decreased during breeding (Vitalini et al. 2011). Lychees treated with melatonin had higher levels of total phenolic, flavonoid, and anthocyanin compounds, which delayed their own postharvest discoloration. While the MT content was highest at the tip of mulberry leaves and the young leaves had a higher content than the old leaves (Pothinuch and Tongchitpakdee 2011), the concentration of MT was higher in bell pepper seedlings but then progressively declined (Xu et al. 2019). The intensity of fruit respiration increases as the "Red Fuji" apple (*Malus domestica* Borkh. cv. Red) ripens and leads to increase in reactive oxygen species this increase in reactive oxygen species matches the trend of change in the associated endogenous melatonin content (Lei et al. 2013). Foreign melatonin is capable of inducing the cell wall-modifying enzymes polygalacturonase (Estravis-Barcala et al. 2020), pectin esterase1 (PE1), β-galactosidase (TBG4), and expansin1 (Exp1), as well as the expression of genes involved in fruit colour development like phytoene synthase1 (PSY1) and carotenoid isomerase (CRTISO), aquaporin genes such as SlPIP12Q, SlPIPQ, SlPIP21Q, and SlPIP22; increased ethylene production; and fruit softening and ripening (Fig. 3.1).

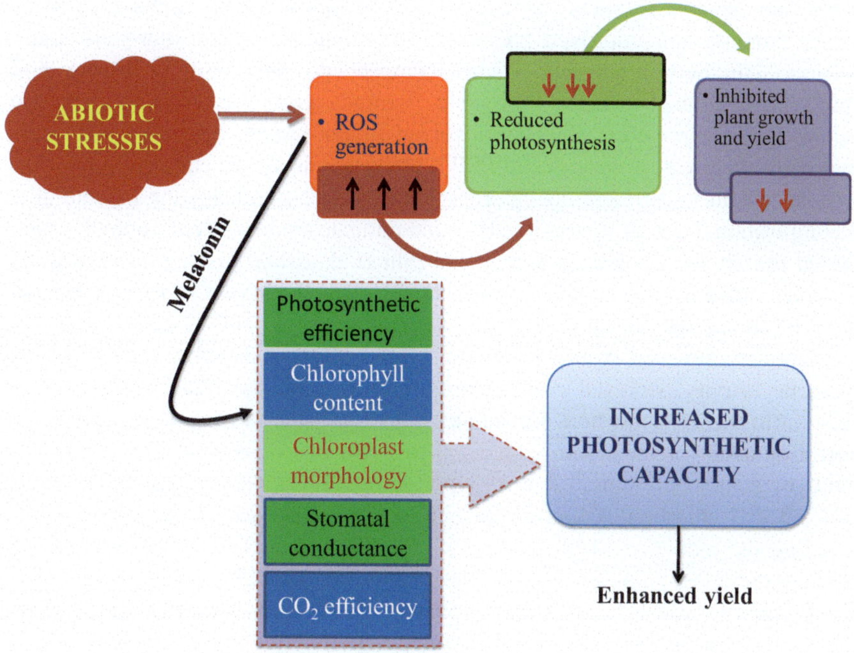

Fig. 3.1 Scheme of melatonin action as positive effector on photosynthetic processes

3.3 Role of Melatonin on Photosynthesis Under Normal Conditions

Melatonin has been found to be effective in plants for increasing chlorophyll's photosynthetic activity while maintaining redox homeostasis, in addition to aiding in the maintenance of chlorophyll (Sharif et al. 2018b). By preventing damage to the photosynthetic apparatus through the elevated amounts of melatonin, it protects many horticultural products, including wheat, barley, sweet cherry, and many others (Sharma and Zheng 2019). To prove this claim, numerous experiments have been conducted. In one trial, melatonin was sprayed on cucumber seedlings between 25 and 100 μm in size. Both seedlings that were under normal circumstances and those that were under heat stress experienced an increase in photosynthetic rate. The conversion of carbohydrates for greater photosynthesis, however, resulted in a decrease in the CO_2 concentration (Xu 2010). Numerous studies on the impact of melatonin on transgenic plants' increased photosynthetic activity, in addition to horticultural products, have lately been published (Byeon and Back 2014; Zheng et al. 2017). One experiment observed a rise in endogenous melatonin production in the chloroplasts of different transgenic A. thaliana lines. In contrast to the wild-type, their chloroplasts displayed an improved redox state with reduced ROS levels when exposed to salt stress. All the transgenic plants were more resistant to salt stress due to the high melatonin content and reduced ROS levels. This was demonstrated by the higher biomass and improved photosynthetic efficiency of the transgenic plants under salt-stressed circumstances when compared to the wild-type (Zheng et al. 2017). In a similar manner, transgenic Bermuda grass treated with melatonin increased the expression of genes involved in photosynthesis under salt stress, leading to the greatest increases in PSII's photochemical efficiency and total chlorophyll content by accelerating chlorophyll biosynthesis and slowing the rate of its decomposition. In order to preserve PSII and mitigate the loss of chlorophyll content caused by salt stress, melatonin is crucial (Shi et al. 2015). In accordance with the aforementioned results, a different study investigated the impact of foliar melatonin spray on maize seedling growth under drought stress. Results revealed that when compared to untreated plants, melatonin-treated plants had higher rates of photosynthetic activity, stomatal conductances, and transpiration (Ye et al. 2016). In addition to the aforementioned example, numerous other studies and experiments were evaluated for melatonin's potential to boost photosynthetic activity. Higher levels of soluble proteins, rubisco, and chlorophyll, as well as greater levels of nitrogen and chlorophyll, can all be induced by melatonin. It has the power to speed up the induction of photosynthesis-related proteins during leaf senescence. Three of the ATP syntheses in chloroplasts can be regulated by melatonin. It plays a part in improving plants' capacity for photosynthetic activity and can increase the effectiveness of photosystems I and II during periods of light and darkness.

3.4 Effect of Melatonin on Photosynthesis Under Stressed Conditions

Photosynthesis is the process in plants which is restricted to the green parts of plants, i.e., leaves (Pan et al. 2012). Numerous metabolic processes are under the supervision of photosynthesis, which transforms light energy into chemical energy (Demmig-Adams et al. 2018). The chloroplast, which is extremely susceptible to abiotic stresses, is the site of both the light and dark reactions of photosynthesis. Under abiotic stress, it has been demonstrated that melatonin application boosted photosynthesis (Han et al. 2017; Ding et al. 2017, Table 3.1).

3.4.1 Under Drought Stress

Melatonin helps plants recover their photosynthetic efficiency by shielding the photosynthetic apparatus from the harmful impacts of drought (Meng et al. 2014; Liang et al. 2018). Melatonin enhances photosynthesis, transpiration, and stomatal conductance and stops the degradation of the chlorophyll molecule under drought stress (Liang et al. 2018; Karaca and Cekic 2019). Enzymes such as chlorophyllase

Table 3.1 Protective role of melatonin on photosynthesis in various crops against different abiotic stresses

Plant	Stress	Photosynthetic improvement	References
Cucumber	Salinity	Enhanced P_N, PSII, and chlorophyll amount	Wang et al. (2016)
Watermelon	Salinity	Alleviated P_N, Gs, SPAD chlorophyll, and qP	Li et al. (2017)
Maize	Drought	Increased the efficiency of PSI and PSII, and boost electron transport	Guo et al. (2020)
Brassica napus	Chromium	Protected PSII and helped out in balancing PSII activity	Ayyaz et al. (2020)
Tomato	High temperature	Increased photosynthetic pigment content. Increases CO_2 assimilation	Jahan et al. (2021)
Coffea arabica	Drought	Knockout the expression of gene *PAO* encoding pheophorbide *a* oxygenase, and upregulated the expression gene *RBCS2*	Cherono et al. (2021)
Solanum lycopersicum	Salinity and cadmium	Enhanced maximum quantum energy (F_V/F_M) and performance index (PI_{ABS})	Bhardwaj et al. (2022)
Capsicum annuum	Cold stress	Rubisco and fructose activity increased with higher efficiency of PSII and PSI	Altaf et al. (2022)
Capsicum annuum	Low temperature and low light	Enhanced expressions of PSII reaction Centre proteins genes as well as increased Fv/Fm, qP, and electron transport rate	Li et al. (2022)
Tomato	Salt	Improved photosynthesis and stomatal behaviour	Khan et al. (2022)
Lentil	Drought	Improved gas exchange parameters and photosynthetic pigments	Yasmeen et al. (2022)

(Chlase), pheophytinase (PPH), and chlorophyll degrading peroxidase (Chl-PRX) cause the breakdown of chlorophyll (Ma et al. 2018). The downregulation of genes such as Chlase, PPH, and Chl-PRX is the cause of the decrease in chlorophyll degradation following melatonin therapy (Ma et al. 2018). Additionally, under drought stress, melatonin restores the content of photosynthetic accessory pigments like carotenoids (Liang et al. 2019). The breakdown of chlorophyll is carried out in part by another enzyme called pheophorbide-a-oxygenase (PAO). Under drought conditions, melatonin slows down the rate of chlorophyll deterioration by downregulating the transcript levels of PAO (Wang et al. 2013). Improved photochemical efficiency (Fv/Fm) of photosystem II (PSII) and a higher photosynthetic electron transport rate (ETR) are associated with increased photosynthetic rate induced by melatonin (Liang et al. 2019). Under drought stress, non-photochemical quenching is intensified and negatively affects photosynthetic efficacy. Application of melatonin to drought-stressed plants, however, aids in the recovery of photosynthetic efficiency (Wang et al. 2013). Another explanation for the improved photosynthetic efficiency under water deficit conditions in melatonin-treated plants is the expansion of the leaf area (Campos et al. 2019). Melatonin shields the chloroplast structures in leaves from oxidative damage, which is the primary cause of the melatonin-mediated improvement of photosynthesis in drought stress (Meng et al. 2014; Cui et al. 2017). Under water deficit conditions, the length of the chloroplast progressively reduces, and the membrane, stromal lamellae, grana, and thylakoids are also disrupted. All these negative impacts of drought on the chloroplast structure are, however, mitigated by melatonin treatment (Meng et al. 2014). After melatonin treatment, leaves under drought stress had improved relative water content, which further favoured the protection of chloroplast structures (Fleta-Soriano et al. 2017). Chlorophyll protection may also be aided by the higher water potential in melatonin-treated plants under drought conditions (Campos et al. 2019). Additionally, after receiving melatonin treatment, drought-stressed plants exhibit better photosynthesis due to an increase in stomata length and a recovery in the shape of palisade tissue along with less damage to spongy tissue cells (Meng et al. 2014). Melatonin also helps to sustain cell turgor, which improves stomatal opening and conductance (Meng et al. 2014). In melatonin-treated plants, this improved stomatal conductance aids in better water and CO_2 transport, which eventually benefits photosynthesis (Cui et al. 2017). The mesophyll cells' controlled water balance and turgor pressure by melatonin further support all of these processes (Cui et al. 2017). Additionally, it has been proposed that melatonin upregulates the transcript levels of genes involved in ABA degradation while concurrently downregulating MdNCED3, a crucial gene in the ABA biosynthetic pathway, in the Malus plant (MdCYP707A1 and MdCYP707A1). The anti-oxidative response and effective H_2O_2 scavenging that accompanied this chemical reaction were also present. Better stomatal function is thought to be the outcome of the synergistic action of both of these mechanisms (Li et al. 2015). Recent research has shown that melatonin regulates the carbon fixation pathway at the molecular level, causing plants growing under drought stress to recover their photosynthetic efficiency (Liang et al. 2019). Melatonin increases the transcript levels of several important carbon fixation pathway enzymes, including RUBISCO (ribulose bisphosphate carboxylase), PGK (phosphoglycerate kinase), GAPA (glyceraldehyde-3-phosphate dehydrogenase), FBA (fructose-bisphosphate aldolase), FBP (fructose-1,6-bisphosphatase), TIM (triosephosphate isomerase), SEBP (sedoheptulose-1,7-bisphosphatase), TKT (transketolase), RPI (ribose 5-phosphate isomerase), and RPK (phosphoribulokinase) (Liang et al. 2019).

3.4.2 Under Salt Stress

Salt stress can have an indirect impact on photosynthesis, a crucial physio-chemical mechanism that produces energy in higher plants (Li et al. 2017; Meloni et al. 2003). Lower levels of photosynthesis are frequently linked to declines in output for many plant species experiencing salt stress (Meloni et al. 2003). Stomatal closure and a compromised photosynthetic apparatus are two potential causes of the salt-induced photosynthesis decrease (Meloni et al. 2003). Stomatal closure can result from salt stress, and one of the metrics used to assess photosynthesis is stomatal conductance (Gs) (Meloni et al. 2003). According to Li et al. (2017), the parameters of chlorophyll fluorescence include maximum PSII photochemical efficiency (Fv/Fm), photochemical quenching (qP), non-photochemical quenching (Y(NPQ)), and true PSII photochemical efficiency (Y(II), among others. Melatonin assists in the control of plant photosynthesis under salt stress in addition to its broad-spectrum antioxidant effects. In watermelons, where salt stress substantially reduced the leaf net photosynthetic rate (Pn), Gs, chlorophyll content, Y(II), and qP, pretreatment with melatonin at various concentrations (50–500 μM) significantly improved salt tolerance. However, preparation with melatonin reduced this decline. By reducing stomatal restriction, melatonin can also safeguard watermelon photosynthesis (Li et al. 2017). Similar outcomes were observed in salt-stressed cucumber seedlings, where exogenous melatonin at concentrations of 50–150 μM greatly increased the photosynthetic capability of the cucumber. Increased PN, PSII's maximal quantum efficiency, and total chlorophyll content are all signs of improved photosynthesis (Wang et al. 2016). Chlorophyll content in radish seedlings rose with melatonin treatment under salt stress, and the 100 M dose was the most effective (Jiang et al. 2017). By slowing the rate at which chlorophyll degrades, melatonin also improved the salinity tolerance of rice seedlings (Liang et al. 2015). Under salt stress, a significant increase in Pn was seen even though the chlorophyll content in melatonin-treated maize seedlings remained the same (Jiang et al. 2016). The protective effects of exogenous melatonin on photosynthesis have also been noted in tomato, apple, and soybean (Li et al. 2012; Wei et al. 2015; Zhou et al. 2016). Overall, exogenous melatonin enhances salt stress endurance by successfully reducing chlorophyll degradation and stomatal closure brought on by salt stress.

3.4.3 Under Heavy Metal Stress

By reducing Chl biosynthesis, carbonic anhydrase (CA) activity, weakening cell wall growth, cell division, and accumulation of lignin and suberin in plants, HM toxicity disrupts the plant photosynthesis system (Masood et al. 2012). According to Al-Huqail et al. (2020), excessive mineral stress decreased the universal biosynthetic precursor compound (δ-ALA) of the photosynthetic pigments and the activity of δ-ALAD while increasing Chl degradation and the activity of Chl degrading enzyme (Chlase) in wheat plants. Additionally, too much HM causes oxidative damage, which can harm reaction centres, change the structure of thylakoids, and cause soft parenchymatic tissue to develop irregularly (Ashraf et al. 2016). Exogenous MT accelerates photosynthesis under HM toxicity by activating the enzymes involved in pigment production and the photosynthetic pathway. According to Al-Huqail et al. (2020), foliar application of MT increased the biosynthesis of photosynthetic pigments by reducing Chl degradation by downregulating Chlase activity and boosting the activity of δ- ALAD, CA, and RuBisco as well as the content of δ-ALA. In order to advance plant photosynthesis while under Cd stress, MT can decrease the rate of Chl degradation and increase Chl concentration (Shi et al. 2015). Based on its involvement in the biosynthesis of porphyrins, glycine, and succinyl-CoA by regulating the activity of δ-aminolevulinate synthase, MT's beneficial contributions to the restoration of altered Chl synthesis and enzyme activity may be tracked (Sarropoulou et al. 2012). Additionally, MT raises ferredoxins, which stop Chl from degrading. Ferredoxins increase the amount of reduced AsA and lower ROS levels, which ultimately shield Chl from degradation by reducing the overproduction of high-energy electrons from the photosynthetic electron transport chain. In tomato seedlings treated with MT and exposed to B toxicity, increased carbonic anhydrase (CA) and RuBisco activity may have improved carbon fixation by maintaining acid-base balance, ion exchange, and a steady supply of CO_2 (Siddiqui et al. 2019), which was replicated in increasing photosynthesis rate. Under vanadium (V) stress, MT treatment significantly improves photosynthesis efficacy in watermelon (Nawaz et al. 2018) and tomato (Lin et al. 2018) by boosting photosynthesis and antioxidant enzyme activities and, in the end, delaying leaf senescence. Furthermore, by applying MT under various stress conditions, the PS II (Fv/Fm) smooth functioning in tomato plants has been noted (Martinez et al. 2018). Melatonin was applied exogenously to canola plants to reduce the toxic effects of chromium on plant development. This was done by altering photosynthesis, improving the efficiency of photosystem II, and controlling the flow of electrons to prevent oxidative damage to the photo-inhibition of PSII (Ayyaz et al. 2020).

3.5 Mechanism of Action

It is unclear how melatonin functions in plants. Transducing signals to the enzyme phospholipase C are known to be involved in signalling pathways in reaction to environmental factors in cells that involve GTP-bound proteins known as G-proteins. Phospholipase C is involved in the transformation of inositol phospholipids, which has an impact on Ca^{2+} channels and the ionic pump, causing Ca^{2+} to be released from intracellular storage systems (vacuoles, mitochondria, and endoplasmic reticulum) and enter the cytoplasm. In addition to Ca^{2+} intracellular signals, environmental factors can also trigger cAMP and cGMP. Melatonin was found to be incorporated into intracellular protein receptors linked to GTP and located in vacuolar membranes in *Gonyaulax polyedra*, which resulted in proton efflux. The first noticeable response of *Gonyaulax polyedra* to melatonin is a 90-fold rise in bioluminescence that occurs at the same time as the release of H^+ into cytoplasm (Hardeland 1993). Cyst development depends on changes in intracellular Ca^{2+} concentration and a drop in intracellular pH. Phosphoinositols also played a role in cyst development, according to research on Gonyaulax polyedra. It has been hypothesized that melatonin causes cyst formation by converting it into 5-metoxytryptamine, which induces vacuolar H^+ -ATPase action and lowers intracellular pH (Tsim et al. 1997). As a result, it appears that the rhythmic rise in melatonin content may be related to signalling pathways that rely on cAMP and Ca^{2+}. The study by Hernandez-Ruiz et al. (2004) verified that melatonin is a plant growth and development regulator with a similar mechanism of action to auxins, though this mechanism is not fully understood. Serotonin and melatonin are two tryptophan derivatives whose role in signalling has been studied, but it is still unclear what that function is (Roshchina 2001). Surprisingly, the most significant signalling tryptophan derivative in plants is auxin, which plays a fundamental regulatory function in plant growth and development. Auxin, which is carried from cell to cell, also resembles neurotransmitters in some ways. According to Baluška et al. (2003), transcellular auxin transport is carried out by a poorly understood vesicle-based mechanism in which putative auxin transporters, or transport facilitators, cycle between the plasma membrane and the endosomes. Auxin and the release of neurotransmitters from neuronal cells appear to share some characteristics (Baluška et al. 2003, 2004, 2005). When viewed from the viewpoint of plant neurobiology, auxin's function can be seen in a new light (Baluška et al. 2004). This is because auxin is known to elicit quick electrical responses when applied extracellularly (Felle et al. 1991; Steffens et al. 2001). According to one theory, auxin molecules released by auxin-enriched vesicles trigger electrical reactions in nearby cells within a short period of time (Steffens et al. 2001). According to Baluška et al. (2003), such electrical activation would be similar to signalling molecules with neurotransmitter-like characteristics. The ABP1 (auxin binding protein)-based signalling pathway or other receptors may be responsible for these quick electrical reactions at the plasma membranes in response to extracellular auxin molecules (Steffens et al. 2001). The auxin-induced responses with a lag-time of several minutes to hours, which are based on auxin receptors and typically involve changes in gene expression, are likely to be distinct from this signalling cascade (Yamagami et al. 2004, via auxin-mediated activation of transcriptional regulators known as auxin response factors) (Parry and Estelle 2006). Additionally, by interacting with cell wall peroxidases, released auxin molecules cause the cell wall to produce ROS (Kawano et al. 2001). In plants, these extremely reactive molecules serve as powerful signalling molecules (Apel and Hirt 2004). Nitric oxide, which plays a variety of functions at neuronal synapses, is additionally closely related to auxin signalling (Pagnussat et al. 2004). It is crucial to continue researching the dynamic signalling capabilities of auxin that is carried between cells; this subject is well suited to plant neurobiology.

3.6 Conclusion

Melatonin is a pleiotropic molecule that has a wide range of functions in plants. It acts as a plant growth stimulator, stress protector, and flowering regulator and improves photosynthesis. Based on the research presented in the text, it is possible to draw the conclusion that melatonin plays a role in reducing the effects of a variety of stressors by regulating antioxidant metabolism and shielding the photosynthetic apparatus. Enhancing crop productivity and addressing the major problem of global food insecurity are made simpler by increased photosynthesis. Agronomically speaking, the application of melatonin at the field level can be beneficial because abiotic stress lowers the quality and output of crops. Future studies are needed to understand the genetic mechanisms and metabolic pathways implicated in melatonin exposure-induced stress recovery. To understand how melatonin interacts with distinct phytohormones in various physiological processes and crosstalk with more signalling molecules, future studies utilizing phytohormone mutants are required.

References

Al-Huqail AA, Khan MN, Ali HM, Siddiqui MH, Al-Huqail AA, AlZuaibr FM, Al-Muwayhi MA, Marraiki N, Al-Humaid LA (2020) Exogenous melatonin mitigates boron toxicity in wheat. Ecotoxicol Environ Saf 201:110822

Altaf MA, Shu H, Hao Y, Mumtaz MA, Lu X, and Wang Z (2022) Melatonin affects the photosynthetic performance of pepper (Capsicum annuum l.) seedlings under cold stress. Antioxidants 11(12):2414.

Apel K, Hirt H (2004) Reactive oxygen species: metabolism, oxidative stress, and signal transduction. Annu Rev Plant Biol 55:373–399

Arah IK, Amaglo H, Kumah EK, Ofori H (2015) Preharvest and postharvest factors affecting the quality and shelf life of harvested tomatoes: a mini review. Int J Agron 2015:478041

Arnao MB, Hernández-Ruiz J (2007) Melatonin promotes adventitious-and lateral root regeneration in etiolated hypocotyls of *Lupinus albus* L. J Pineal Res 42(2):147–152

Arnao MB, Hernández-Ruiz J (2009) Protective effect of melatonin against chlorophyll degradation during the senescence of barley leaves. J Pineal Res 46(1):58–63

Arnao MB, Hernández-Ruiz J (2013) Growth conditions determine different melatonin levels in *Lupinus albus* L. J Pineal Res 55(2):149–155

Arnao MB, Hernández-Ruiz J (2015) Functions of melatonin in plants: a review. J Pineal Res 59(2): 133–150

Arnao MB, Hernández-Ruiz J (2018) Melatonin and its relationship to plant hormones. Ann Bot 121(2):195–207

Ashraf MY, Roohi M, Iqbal Z, Ashraf M, Öztürk M, Gücel S (2016) Cadmium (Cd) and lead (Pb) induced changes in growth, some biochemical attributes, and mineral accumulation in two cultivars of mung bean [*Vigna radiata* (L.) Wilczek]. Commun Soil Sci Plant Anal 47(4): 405–413

Ayyaz A, Amir M, Umer S, Iqbal M, Bano H, Gul HS, Noor Y, Javed M, Athar HR, Zafar ZU, Farooq MA (2020) Melatonin induced changes in photosynthetic efficiency as probed by OJIP associated with improved chromium stress tolerance in canola (*Brassica napus* L.). Heliyon 6(7):e04364

Bajwa VS, Shukla MR, Sherif SM, Murch SJ, Saxena PK (2014) Role of melatonin in alleviating cold stress in *Arabidopsis thaliana*. J Pineal Res 56(3):238–245

Baluška F, Šamaj J, Menzel D (2003) Polar transport of auxin: carrier-mediated flux across the plasma membrane or neurotransmitter-like secretion? Trends Cell Biol 13(6):282–285

Baluška F, Mancuso S, Volkmann D, Barlow P (2004) Root apices as plant command centres: the unique 'brain-like' status of the root apex transition zone. Biologia 59(Suppl 13):7–19

Baluška F, Volkmann D, Menzel D (2005) Plant synapses: actin-based domains for cell-to-cell communication. Trends Plant Sci 10(3):106–111

Bhardwaj K, Raina M, Sanfratello GM, Pandey P, Singh A, Rajwanshi R, Negi NP, Rustagi A, Kumar D (2022) Exogenous melatonin counteracts salinity and cadmium stress via photosynthetic machinery and antioxidant modulation in *Solanum lycopersicum* L. J Plant Growth Regul 42:1–17

Byeon Y, Back K (2014) An increase in melatonin in transgenic rice causes pleiotropic phenotypes, including enhanced seedling growth, delayed flowering, and low grain yield. J Pineal Res 56(4): 408–414

Campos CN, Ávila RG, de Souza KRD, Azevedo LM, Alves JD (2019) Melatonin reduces oxidative stress and promotes drought tolerance in young *Coffea arabica* L. plants. Agric Water Manag 211:37–47

Cao S, Shao J, Shi L, Xu L, Shen Z, Chen W, Yang Z (2018) Melatonin increases chilling tolerance in postharvest peach fruit by alleviating oxidative damage. Sci Rep 8(1):806

Chan Z, Shi H (2015) Improved abiotic stress tolerance of bermudagrass by exogenous small molecules. Plant Signal Behav 10(3):e991577

Chen L, Fan J, Hu Z, Huang X, Amombo E, Liu A, Bi A, Chen K, Xie Y, Fu J (2017) Melatonin is involved in regulation of bermudagrass growth and development and response to low K+ stress. Front Plant Sci 8:2038

Cherono S, Ntini C, Wassie M, Mollah MD, Belal MA, Ogutu C, Han Y (2021) Exogenous application of melatonin improves drought tolerance in coffee by regulating photosynthetic efficiency and oxidative damage. J Am Soc Hort Sci 146(1):24–32

Cui G, Zhao X, Liu S, Sun F, Zhang C, Xi Y (2017) Beneficial effects of melatonin in overcoming drought stress in wheat seedlings. Plant Physiol Biochem 118:138–149

Demmig-Adams B, Stewart JJ, Baker CR, Adams WW III (2018) Optimization of photosynthetic productivity in contrasting environments by regulons controlling plant form and function. Int J Mol Sci 19(3):872

Ding F, Wang M, Liu B, Zhang S (2017) Exogenous melatonin mitigates photoinhibition by accelerating non-photochemical quenching in tomato seedlings exposed to moderate light during chilling. Front Plant Sci 8:244

Estravis-Barcala M, Mattera MG, Soliani C, Bellora N, Opgenoorth L, Heer K, Arana MV (2020) Molecular bases of responses to abiotic stress in trees. J Exp Bot 71(13):3765–3779

Felle H, Peters W, Palme K (1991) The electrical response of maize to auxins. Biochim Biophys Acta Biomembr 1064(2):199–204

Fleta-Soriano E, Díaz L, Bonet E, Munné-Bosch S (2017) Melatonin may exert a protective role against drought stress in maize. J Agron Crop Sci 203(4):286–294

Guo YY, Li HJ, Liu J, Bai YW, Xue JQ, Zhang RH (2020) Melatonin alleviates drought-induced damage of photosynthetic apparatus in maize seedlings. Russ J Plant Physiol 67:312–322

Han QH, Huang B, Ding CB, Zhang ZW, Chen YE, Hu C, Zhou LJ, Huang Y, Liao JQ, Yuan S, Yuan M (2017) Effects of melatonin on anti-oxidative systems and photosystem II in cold-stressed rice seedlings. Front Plant Sci 8:785

Hardeland R (1993) The presence and function of melatonin and structurally related indoleamines in a dinoflagellate, and a hypothesis on the evolutionary significance of these tryptophan metabolites in unicellulars. Experientia 49:614–622

Hardeland R (2013) Melatonin and the theories of aging: a critical appraisal of melatonin's role in antiaging mechanisms. J Pineal Res 55(4):325–356

Hernández-Ruiz J, Arnao MB (2008) Melatonin stimulates the expansion of etiolated lupin cotyledons. Plant Growth Regul 55:29–34

Hernandez-Ruiz J, Cano A, Arnao MB (2004) Melatonin: a growth-stimulating compound present in lupin tissues. Planta, 220:140–144.

Hu W, Tie W, Ou W, Yan Y, Kong H, Zuo J, Ding X, Ding Z, Liu Y, Wu C, Guo Y (2018) Crosstalk between calcium and melatonin affects postharvest physiological deterioration and quality loss in cassava. Postharvest Biol Technol 140:42–49

Jahan MS, Guo S, Sun J, Shu S, Wang Y, Abou El-Yazied A, Alabdallah NM, Hikal M, Mohamed MH, Ibrahim MF, Hasan MM (2021) Melatonin-mediated photosynthetic performance of tomato seedlings under high-temperature stress. Plant Physiol Biochem 167:309–320

Jiang C, Cui Q, Feng K, Xu D, Li C, Zheng Q (2016) Melatonin improves antioxidant capacity and ion homeostasis and enhances salt tolerance in maize seedlings. Acta Physiol Plant 38:82

Jiang Y, Liang D, and Lin L (2017) Effects of melatonin on the growth of radish Seedlings under salt stress. In 4th International Conference on Renewable Energy and Environmental Technology (ICREET 2016) (pp. 383–386). Atlantis Press.

Karaca P, Cekic FÖ (2019) Exogenous melatonin-stimulated defense responses in tomato plants treated with polyethylene glycol. Int J Veg Sci 25(6):601–609

Kawano T, Kawano N, Hosoya H, Lapeyrie F (2001) Fungal auxin antagonist hypaphorine competitively inhibits indole-3-acetic acid-dependent superoxide generation by horseradish peroxidase. Biochem Biophys Res Commun 288(3):546–551

Khan TA, Saleem M, Fariduddin Q (2022) Melatonin influences stomatal behavior, root morphology, cell viability, photosynthetic responses, fruit yield, and fruit quality of tomato plants exposed to salt stress. J Plant Growth Regul 42:1–25

Kolář J, Macháčková I, Eder J, Prinsen E, Van Dongen W, Van Onckelen H, Illnerová H (1997) Melatonin: occurrence and daily rhythm in *Chenopodium rubrum*. Phytochemistry 44(8): 1407–1413

Kolář J, Johnson CH, Macháčková I (2003) Exogenously applied melatonin (N-acetyl-5-methoxytryptamine) affects flowering of the short-day plant *Chenopodium rubrum*. Physiol Plant 118(4):605–612

Lee HY, Byeon Y, Tan DX, Reiter RJ, Back K (2015) Arabidopsis serotonin N-acetyltransferase knockout mutant plants exhibit decreased melatonin and salicylic acid levels resulting in susceptibility to an avirulent pathogen. J Pineal Res 58(3):291–299

Lei Q, Wang L, Tan DX, Zhao Y, Zheng XD, Chen H, Li QT, Zuo BX, Kong J (2013) Identification of genes for melatonin synthetic enzymes in 'Red Fuji' apple (*Malus domestica* Borkh. cv. Red) and their expression and melatonin production during fruit development. J Pineal Res 55(4): 443–451

Lerner AB, Case JD (1959) Part III: general considerations of skin pigmentation: pigment cell regulatory factors. J Investig Dermatol 32(2):211–221

Li C, Wang P, Wei Z, Liang D, Liu C, Yin L, Jia D, Fu M, Ma F (2012) The mitigation effects of exogenous melatonin on salinity-induced stress in *Malus hupehensis*. J Pineal Res 53(3): 298–306

Li C, Tan DX, Liang D, Chang C, Jia D, Ma F (2015) Melatonin mediates the regulation of ABA metabolism, free-radical scavenging, and stomatal behaviour in two malus species under drought stress. J Exp Bot 66(3):669–680

Li H, Chang J, Chen H, Wang Z, Gu X, Wei C, Zhang Y, Ma J, Yang J, Zhang X (2017) Exogenous melatonin confers salt stress tolerance to watermelon by improving photosynthesis and redox homeostasis. Front Plant Sci 8:295

Li J, Ding D, Li N, Xie J, Yu J, Lyv J, Bakpa EP, Zhang J, Wang C, Zhang J (2022) Melatonin enhances the low-temperature combined low-light tolerance of pepper (*Capsicum annuum* L.) seedlings by regulating photosynthesis, carotenoid, and hormone metabolism. Environ Exp Bot 199:104868

Liang C, Zheng G, Li W, Wang Y, Hu B, Wang H, Wu H, Qian Y, Zhu XG, Tan DX, Chen SY (2015) Melatonin delays leaf senescence and enhances salt stress tolerance in rice. J Pineal Res 59(1):91–101

Liang B, Ma C, Zhang Z, Wei Z, Gao T, Zhao Q, Ma F, Li C (2018) Long-term exogenous application of melatonin improves nutrient uptake fluxes in apple plants under moderate drought stress. Environ Exp Bot 155:650–661

Liang D, Ni Z, Xia H, Xie Y, Lv X, Wang J, Lin L, Deng Q, Luo X (2019) Exogenous melatonin promotes biomass accumulation and photosynthesis of kiwifruit seedlings under drought stress. Sci Hortic 246:34–43

Lin L, Li J, Chen F, Liao MA, Tang Y, Liang D, Xia H, Lai Y, Wang X, Chen C, Ren W (2018) Effects of melatonin on the growth and cadmium characteristics of *Cyphomandra betacea* seedlings. Environ Monit Assess 190:1–8

Liu J, Zhang R, Sun Y, Liu Z, Jin W, Sun Y (2016) The beneficial effects of exogenous melatonin on tomato fruit properties. Sci Hortic 207:14–20

Ma X, Zhang J, Burgess P, Rossi S, Huang B (2018) Interactive effects of melatonin and cytokinin on alleviating drought-induced leaf senescence in creeping bentgrass (*Agrostis stolonifera*). Environ Exp Bot 145:1–11

Martinez V, Nieves-Cordones M, Lopez-Delacalle M, Rodenas R, Mestre TC, Garcia-Sanchez F, Rubio F, Nortes PA, Mittler R, Rivero RM (2018) Tolerance to stress combination in tomato plants: new insights in the protective role of melatonin. Molecules 23(3):535

Masood A, Iqbal N, Khan NA (2012) Role of ethylene in alleviation of cadmium-induced photosynthetic capacity inhibition by Sulphur in mustard. Plant Cell Environ 35(3):524–533

Meloni DA, Oliva MA, Martinez CA, Cambraia J (2003) Photosynthesis and activity of superoxide dismutase, peroxidase and glutathione reductase in cotton under salt stress. Environ Exp Bot 49(1):69–76

Meng JF, Xu TF, Wang ZZ, Fang YL, Xi ZM, Zhang ZW (2014) The ameliorative effects of exogenous melatonin on grape cuttings under water-deficient stress: antioxidant metabolites, leaf anatomy, and chloroplast morphology. J Pineal Res 57(2):200–212

Murch SJ, Alan AR, Cao J, Saxena PK (2009) Melatonin and serotonin in flowers and fruits of *Datura metel* L. J Pineal Res 47(3):277–283

Murch SJ, Hall BA, Le CH, Saxena PK (2010) Changes in the levels of indoleamine phytochemicals during véraison and ripening of wine grapes. J Pineal Res 49(1):95–100

Nawaz MA, Huang Y, Bie Z, Ahmed W, Reiter RJ, Niu M, Hameed S (2016) Melatonin: current status and future perspectives in plant science. Front Plant Sci 6:1230

Nawaz MA, Jiao Y, Chen C, Shireen F, Zheng Z, Imtiaz M, Bie Z, Huang Y (2018) Melatonin pretreatment improves vanadium stress tolerance of watermelon seedlings by reducing vanadium concentration in the leaves and regulating melatonin biosynthesis and antioxidant-related gene expression. J Plant Physiol 220:115–127

Pagnussat GC, Lanteri ML, Lombardo MC, Lamattina L (2004) Nitric oxide mediates the indole acetic acid induction activation of a mitogen-activated protein kinase cascade involved in adventitious root development. Plant Physiol 135(1):279–286

Pan J, Lin S, Woodbury NW (2012) Bacteriochlorophyll excited-state quenching pathways in bacterial reaction centers with the primary donor oxidized. J Phys Chem B 116(6):2014–2022

Park S, Le TNN, Byeon Y, Kim YS, Back K (2013) Transient induction of melatonin biosynthesis in rice (*Oryza sativa* L.) during the reproductive stage. J Pineal Res 55(1):40–45

Parry G, Estelle M (2006) Auxin receptors: a new role for F-box proteins. Curr Opin Cell Biol 18(2):152–156

Pothinuch P, Tongchitpakdee S (2011) Melatonin contents in mulberry (*Morus* spp.) leaves: effects of sample preparation, cultivar, leaf age and tea processing. Food Chem 128(2):415–419

Qian Y, Tan DX, Reiter RJ, Shi H (2015) Comparative metabolomic analysis highlights the involvement of sugars and glycerol in melatonin-mediated innate immunity against bacterial pathogen in Arabidopsis. Sci Rep 5(1):15815

Reiter RJ (1991) Pineal melatonin: cell biology of its synthesis and of its physiological interactions. Endocr Rev 12(2):151–180

Reiter RJ, Tan DX, Zhou Z, Cruz MHC, Fuentes-Broto L, Galano A (2015) Phytomelatonin: assisting plants to survive and thrive. Molecules 20(4):7396–7437

Roshchina VV (2001) Neurotransmitters in plant life. CRC, Boca Raton, FL

Sarropoulou V, Dimassi-Theriou K, Therios I, Koukourikou-Petridou M (2012) Melatonin enhances root regeneration, photosynthetic pigments, biomass, total carbohydrates and proline content in the cherry rootstock PHL-C (*Prunus avium* × *Prunus cerasus*). Plant Physiol Biochem 61:162–168

Sharif R, Mujtaba M, Ur Rahman M, Shalmani A, Ahmad H, Anwar T, Tianchan D, Wang X (2018a) The multifunctional role of chitosan in horticultural crops; a review. Molecules 23(4): 872

Sharif R, Xie C, Zhang H, Arnao MB, Ali M, Ali Q, Muhammad I, Shalmani A, Nawaz MA, Chen P, Li Y (2018b) Melatonin and its effects on plant systems. Molecules 23(9):2352

Sharma A, Zheng B (2019) Melatonin mediated regulation of drought stress: physiological and molecular aspects. Plan Theory 8(7):190

Shi H, Jiang C, Ye T, Tan DX, Reiter RJ, Zhang H, Liu R, Chan Z (2015) Comparative physiological, metabolomic, and transcriptomic analyses reveal mechanisms of improved abiotic stress resistance in bermudagrass [*Cynodon dactylon* (L). Pers.] by exogenous melatonin. J Exp Bot 66(3):681–694

Siddiqui MH, Alamri S, Alsubaie QD, Ali HM, Ibrahim AA, Alsadon A (2019) Potential roles of melatonin and sulfur in alleviation of lanthanum toxicity in tomato seedlings. Ecotoxicol Environ Saf 180:656–667

Steffens B, Feckler C, Palme K, Christian M, Böttger M, Lüthen H (2001) The auxin signal for protoplast swelling is perceived by extracellular ABP1. Plant J 27(6):591–599

Sun Q, Zhang N, Wang J, Zhang H, Li D, Shi J, Li R, Weeda S, Zhao B, Ren S, Guo YD (2015) Melatonin promotes ripening and improves quality of tomato fruit during postharvest life. J Exp Bot 66(3):657–668

Sun W, Wang Z, Cao J, Cui H, Ma Z (2016) Cold stress increases reactive oxygen species formation via TRPA1 activation in A549 cells. Cell Stress Chaperones 21:367–372

Tan DX, Xu B, Zhou X, Reiter RJ (2018) Pineal calcification, melatonin production, aging, associated health consequences and rejuvenation of the pineal gland. Molecules 23(2):301

Tsim ST, Wong JT, Wong YH (1997) Calcium ion dependency and the role of inositol phosphates in melatonin-induced encystment of dinoflagellates. J Cell Sci 110(12):1387–1393

Vitalini S, Gardana C, Zanzotto A, Simonetti P, Faoro F, Fico G, Iriti M (2011) The presence of melatonin in grapevine (*Vitis vinifera* L.) berry tissues. J Pineal Res 51(3):331–337

Wang P, Yin L, Liang D, Li C, Ma F, Yue Z (2012) Delayed senescence of apple leaves by exogenous melatonin treatment: toward regulating the ascorbate–glutathione cycle. J Pineal Res 53(1):11–20

Wang P, Sun X, Li C, Wei Z, Liang D, Ma F (2013) Long-term exogenous application of melatonin delays drought-induced leaf senescence in apple. J Pineal Res 54(3):292–302

Wang L, Zhao Y, Reiter RJ, He C, Liu G, Lei Q, Zuo B, Zheng XD, Li Q, Kong J (2014) Changes in melatonin levels in transgenic 'micro-Tom' tomato overexpressing ovine AANAT and ovine HIOMT genes. J Pineal Res 56(2):134–142

Wang LY, Liu JL, Wang WX, Sun Y (2016) Exogenous melatonin improves growth and photosynthetic capacity of cucumber under salinity-induced stress. Photosynthetica 54:19–27

Wei W, Li QT, Chu YN, Reiter RJ, Yu XM, Zhu DH, Zhang WK, Ma B, Lin Q, Zhang JS, Chen SY (2015) Melatonin enhances plant growth and abiotic stress tolerance in soybean plants. J Exp Bot 66(3):695–707

Xu XD (2010) Effect of exogenous melatonin on the physiological response of cucumber seedlings under high temperature stress. Doctoral dissertation, Master's thesis. Northwest Agriculture and Forestry University, Xianyang, China

Xu T, Chen Y, Kang H (2019) Melatonin is a potential target for improving post-harvest preservation of fruits and vegetables. Front Plant Sci 10:1388

Yamagami M, Haga K, Napier RM, Iino M (2004) Two distinct signaling pathways participate in auxin-induced swelling of pea epidermal protoplasts. Plant Physiol 134(2):735–747

Yasmeen S, Wahab A, Saleem MH, Ali B, Qureshi KA, Jaremko M (2022) Melatonin as a foliar application and adaptation in lentil (*Lens culinaris* Medik.) crops under drought stress. Sustainability 14(24):16345

Ye J, Wang S, Deng X, Yin L, Xiong B, Wang X (2016) Melatonin increased maize (*Zea mays* L.) seedling drought tolerance by alleviating drought-induced photosynthetic inhibition and oxidative damage. Acta Physiol Plant 38(2):48

Zhang N, Zhang HJ, Zhao B, Sun QQ, Cao YY, Li R, Wu XX, Weeda S, Li L, Ren S, Reiter RJ (2014) The RNA-seq approach to discriminate gene expression profiles in response to melatonin on cucumber lateral root formation. J Pineal Res 56(1):39–50

Zhang N, Sun Q, Li H, Li X, Cao Y, Zhang H, Li S, Zhang L, Qi Y, Ren S, Zhao B (2016) Melatonin improved anthocyanin accumulation by regulating gene expressions and resulted in high reactive oxygen species scavenging capacity in cabbage. Front Plant Sci 7:197

Zhang L, Zhu M, Ren L, Li A, Chen G, Hu Z (2018) The SlFSR gene controls fruit shelf-life in tomato. J Exp Bot 69(12):2897–2909

Zheng X, Tan DX, Allan AC, Zuo B, Zhao Y, Reiter RJ, Wang L, Wang Z, Guo Y, Zhou J, Shan D (2017) Chloroplastic biosynthesis of melatonin and its involvement in protection of plants from salt stress. Sci Rep 7(1):1–12

Zhou X, Zhao H, Cao K, Hu L, Du T, Baluška F, Zou Z (2016) Beneficial roles of melatonin on redox regulation of photosynthetic electron transport and synthesis of D1 protein in tomato seedlings under salt stress. Front Plant Sci 7:1823

Chapter 4
Melatonin and Its Crosstalk with Other Plant Hormones Under Stressful Environments

Mohammad Faizan, Nazish, Ahmad Faraz, S. Maqbool Ahmed, and Shamsul Hayat

Abstract The discovery of melatonin in plants was started in the late 90s, but its functions, signaling, and crosstalk with various phytohormones is still unknown. The research on melatonin in plants has increased spectacularly during recent years and the function of this assumed plant hormone under stress (biotic and abiotic) conditions has also been accounted. In the present chapter, we talk about the key job of melatonin in the morphological and physiological functioning of the plants, its crosstalk with auxin, gibberellic acid, cytokinin, abscisic acid, ethylene, jasmonic acid, and salicylic acid under abiotic stress as well as normal environmental conditions. All these features of melatonin recommend that phyto-melatonin is an important factor in agriculture for crop progress even under stress conditions.

Keywords Melatonin · Abiotic stress · Gibberellic acid · Salicylic acid

4.1 Introduction

Melatonin (*N*-acetyl- 5-methoxytryptamine) is most prevalent molecule among plants, animals, and microorganisms and considered as an indolic compound. Melatonin has structural similarities with other crucial compounds such as

M. Faizan (✉) · S. M. Ahmed
Botany Section, School of Sciences, Maulana Azad National Urdu University, Hyderabad, Telangana, India
e-mail: mohammadfaizan@manuu.edu.in

Nazish
Department of Science, JMB Institute of Life Sciences and Higher Education, Pilibhit, India

A. Faraz
Department of Biotechnology, School of Life Sciences, Glocal University, Saharanpur, India

S. Hayat
Plant Physiology Section, Department of Botany, Aligarh Muslim University, Aligarh, India

A. Sharma, G. J. Ahammed (eds.), *Melatonin in Plants: Role in Plant Growth, Development, and Stress Response*, Plant Life and Environment Dynamics,
https://doi.org/10.1007/978-981-99-8051-2_4

tryptophan, serotonin, and indole-3-acetic acid (IAA). The presence of melatonin in plants revealed that melatonin is a common and multifunctional metabolite among the plant world. All plant parts including the root, stems, leaves, flowers, fruits, and seeds found to have melatonin (Fan et al. 2018; Wang et al. 2018). The potential roles of melatonin in plants are its effect on abiotic and biotic stress caused by salinity, drought, extreme temperatures, heavy metals, and UV radiation (Arnao and Hernández-Ruiz 2017). During biotic and abiotic stress, it regulates plant growth and vegetative development such as rooting, photosynthetic efficiency, leaf senescence, and biomass yield along with regulating the flowering and ripening of fruits (Nawaz et al. 2016). In stress conditions melatonin acts as a strong free radical scavenger against harmful reactive molecules like reactive oxygen species and reactive nitrogen species (ROS/RNS). However, the role of melatonin as a natural antioxidant adverse to ROS/RNS and the lack of pro-oxidant effects have been the matter of a great interest of research (Reiter et al. 2014). In the present article, the interaction between melatonin and plant growth regulators such as auxin, gibberellins (GA), cytokinins (CKs), ethylene, jasmonic acid (JA), abscisic acid (ABA), and salicylic acid (SA) are discussed.

4.2 Interplay Between Melatonin and Other Phytohormones, and the Underlying Signaling Events

4.2.1 Auxin and Melatonin

Melatonin found to have potential roles in plant growth and development mainly in root architecture. As melatonin and auxin have similar chemical structure as well as biosynthetic pathway. The signaling pathways of these two compounds also have a potential linkage (Park 2011). Various studies have been suggested that melatonin can promote plant growth like auxin in different plants such as wheat, barley, rice, tomato, soya been, cucumber and in *Arabidopsis thaliana* as a model organism. Several reports reveal that mutations of the flavonoid pathway in tomato and *Arabidopsis* play pivotal role on lateral root development by affecting transport of auxin. They observe the relationship between melatonin and auxin with flavonoid pathway and comprise the collaborative effect of melatonin and auxin between wild type Columbia and their H_4, H_5 mutants. Furthermore, the synergistic effect of melatonin and auxin found to stimulate lateral root development in wild type *Arabidopsis* but such collaborative effect was not observed in knockout mutants of auxin transport related genes. Therefore, melatonin promotes the root development through the regulation of auxin distribution via transformation of auxin transport (Ren et al. 2019).

The melatonin was found to have growth-promoting activity that is more specific to auxin-like roles. Melatonin has been observed to persuade growth in roots and aerial parts, e.g., *Avena, Hordeum, Triticum, Lupinus, Arabidopsis, Oryza,*

Helianthus, Punica, Prunus, and *Brassica* as well as in tomato, maize, and soybean plants (Arnao and Hernández-Ruiz 2017). However, the high concentration (>10 μM) of melatonin inhibits the growth in roots. The growth-promoting activity of melatonin is increases in stressful environment, as in case of *Zea mays* (Kim et al. 2016) and *Helianthus* (Mukherjee et al. 2014) grown in salt stress, in *Arabidopsis* and *Cynodon* exposed to cold stress (Bajwa et al. 2014; Shi et al. 2015a, b, c), and in *Malus* grown under salt stress or nutrient-deficiency. In several reports, melatonin treatment found to induce endogenous IAA level as compared to untreated plants as noticed in tomato and *Brassica juncea* (Wen et al. 2016). However, in genetically modified plants where the overproduction of melatonin occurs, a considerable reduction in IAA levels has also been observed. For instance, the accumulation of endogenous melatonin increase up to six times in tomato over expressing sheep SNAT/HIOMT and *Arabidopsis* over expressing apple HIOMT that was higher than that of wild plants, while the endogenous IAA level decreased sevenfold in tomato and 1.4-fold in *Arabidopsis*. The phenotypic effects of these transgenic plants found to be similar with auxin-like responses, elevating root growth, and rhizogenesis, along with reduced apical dominance in tomato overexpressing sheep SNAT/ASMT and a variety of lateral roots in transgenic *Arabidopsis* (Zuo et al. 2014). These observations suggest that melatonin and IAA both performs similar physiological functions. During a recent study in *Arabidopsis*, root meristem size found to reduce by melatonin treatment that ultimately decreases the endogenous IAA content. Wang et al. (2016) observed that melatonin regulates the root meristem size through the suppression of auxin synthesis and transport. Moreover, upregulation and downregulation of different auxin related transcription factors were exhibited in rice and *Arabidopsis* under the influence of melatonin treatment (Liang et al. 2017). However, IAA, but not melatonin, can operate the auxin-stimulating gene expression marker DR5:GUS in *Arabidopsis* (Koyama et al. 2013). These finding indicates that melatonin can act analogous to IAA in the induction of both lateral and adventitious root development.

In biosynthesis of IAA, the transcription levels of some crucial genes remarkably declined in response to melatonin treatment like those encoding TAA and YUCCA family proteins (Arnao and Hernández Ruiz 2018). Melatonin seems to induce rhizogenesis without auxin, the enhanced expression of the auxin inducible DR5: GUS reporter was not observed in melatonin-treated *Arabidopsis* (Koyama et al. 2013). The expression of various auxin related genes was also remains similar in cucumber roots before and after the melatonin treatment (Zhang et al. 2014). Exclusively genes encoding several auxin influx carrier proteins (AUX1/LAX), regulates the lateral root development, root hair development and gravitropism were critically down regulated in *Arabidopsis* with regards to melatonin application (Weeda et al. 2014). The expression level of many auxin-dependent transcription factors such as NAC, MYB, HDZIP, WRKY, and bHLH was also altered by the melatonin treatment (Liang et al. 2017). On the basis of these exploratory observations, it was concluded that melatonin plays a significant role in the regulation of root morphology through the activation of auxin signaling pathway and possibly influence the polar auxin transport and perception. Research indicates that melatonin as well as auxin signals are mediated via various signaling cascades, which eventually merge at some checkpoints. An emerging potential player in melatonin and auxin signaling is nitric oxide (NO), a signaling molecule having myriad roles in regulating growth, development, and environmental interactions in plants (Sun et al. 2014; Sanz et al. 2015). NO modulate auxin responses, those results in adventitious root development and also found to regulate lateral roots and root hair development along with root gravitropism (Yu et al. 2014). In *Arabidopsis* the elevated nitric oxide (NO) level could be retarded auxin transport and auxin responses across a PIN1 dependent mechanism, consequently suppressing the root meristem activity (Fernández-Marcos et al. 2011). Melatonin treatment in tomato triggered adventitious root formation through the activation of NO production, thereby regulating the expression of a number of auxin-dependent genes, including those associated with auxin transport, signal transduction, and auxin accumulation (Wang et al. 2016). NO treatment also enhanced melatonin production in tomato seedlings, implying to a possible feedback loop (Arnao and Hernández Ruiz 2018).

4.2.2 Gibberellic Acid and Melatonin

In rice seedlings the melatonin synthesis was sufficiently persuade upon exogenous gibberellins 3 (GA_3) treatments, but it was critically declined by paclobutrazol which act as a GA inhibitor. The GA-mediated enhancement of melatonin was found closely linked with enhanced expression levels of melatonin anabolic genes such as ASMT1, T5H, and TDC3 as well as reduced expression levels of some catabolic genes M2H and ASDAC (Hwang and Back 2022). The treatment of immature rice seeds in a paddy field, with exogenous GA application leads to enhanced melatonin production in rice seeds. Different transgenic rice plants down regulating a GA anabolic gene (GA3ox2) and a signaling gene (Gα) exhibited severely declined melatonin levels, revealing in vivo genetic evidence that GA has a positive correlation with melatonin synthesis. GA treatment can also be used to produce melatonin-rich seeds, fruits, and vegetables useful for human health.

Exogenous application of melatonin enhanced the transcript accumulation of GA-synthesis genes GA20ox and GA3ox in cucumber plants in order to support the germination obstruct by salt (Zhang et al. 2014). Several research studies on cucumber and red kraut reveal that melatonin possesses germination elevating properties (Posmyk et al. 2009). During salt stress in *Brassica napus* L. melatonin therapy stimulates the growth of seedling by enhancing GA levels and upregulating three potential GA producing enzymes (GA3ox, GA2ox, and GA20ox). Moreover, the GID genes encoding soluble GA receptor linked with GA and DELLA proteins in order to create a complex that prevents DELLAs from the suppression of GA signaling were also upregulated (Tan et al. 2019). Similar effects were also observed in rice seedlings grown in fluoride-rich soil. The GA concentration and cyclin synthesis in apple plants was found to improve by melatonin treatment (Mao et al. 2020). Melatonin treatment sink the floral transition in *Arabidopsis* by stimulating the expression of flowering locus C and suppressing the expression of flowering locus T (Shi et al. 2015a, b, c). However, the impact of melatonin on GA levels is disputable because it has been observed to increase GA levels in various stress responses.

4.2.3 *Cytokinin and Melatonin*

The exogenous melatonin treatment enhanced cytokinin levels by upregulating LpOG1, LpIPT2, and key CK-biosynthesis-related genes during heat stress whereas under non-stress conditions, cytokinin level remain unchanged by melatonin treatment. These observations showed a positive correlation between endogenous melatonin level and cytokinin. Furthermore, melatonin was also altered cytokinin signaling pathway via the regulation of A-ARRs and B-ARRs, transcription factors associated with CK signaling events (Zhang et al. 2017). *Agrostis stolonifera* (creeping bent grass) efficiently defended from drought-mediated harmful effects such as leaf senescence, decreased growth, and membrane damage through melatonin treatment via increased chlorophyll production, higher antioxidant enzymes activity, increased expression of genes involved in chlorophyll as well as cytokinin production (Ma et al. 2018). In barley leaves melatonin delayed dark-induced senescence through a concentration-dependent manner, and decreased chlorophyll drop in detached leaves. Hence, melatonin could inhibits the induced-senescence process (Arnao and Hernández-Ruiz 2009). The senescence process was also found to inhibit in leaf sections treated with different kinetin (a synthetic cytokinin) concentrations and it also induced the lower chlorophyll loss in treated leaves as compared to control leaves. The kinetin and melatonin combinations obstruct chlorophyll losses when melatonin was used in the concentration of 0.1 and 1 mm (Arnao and Hernández Ruiz 2018). However, CKs cannot stop senescence completely but considerably delay by stimulating nutrient mobilization.

The above demonstrations indicate that the mitigation of heat stress through melatonin is the consequence of an interesting crosstalk between melatonin and cytokinin. Melatonin and cytokinin have recently been found associated with the ripening of sweet cherries (Tijero et al. 2019). In a recent study on *Arabidopsis* melatonin observed to controls primary root growth in collaboration with cytokinin partially via auxin-dependent pathway (Wang et al. 2022).

4.2.4 *Abscisic Acid and Melatonin*

Melatonin can be act through Abscisic acid (ABA) signaling pathways, regulating the salinity and drought stress responses (Sharma and Zheng 2019). The interaction of melatonin with ABA results in a decline of ABA levels as well as the downregulation of ABA biosynthesis enzymes, with reduced sensitivity to ABA in consequence of altered ABA signaling element control. The downregulation of enzyme responsible for ABA biosynthesis, i.e., 9-cis-epoxycarotenoid dioxygenase (NCED) and the upregulation of ABA catabolism genes (CYP707 monooxygenases) affected by melatonin treatment results in instant drop of ABA level. These effects of melatonin and ABA interaction could be observed in Chinese cabbage, cucumbers, apples, and hickory (Arnao and Hernández-Ruiz 2021). A current study on melatonin-treated mango exhibited delayed softening and ripening due to the reduced amounts of ABA and ethylene through the downregulation of ACS, NCED, and ACO, along with pectin-modulating enzymes (Liu et al. 2020). However during water-stress in maize, exogenous application of melatonin is unaffected on ABA levels (Jia and Zhang 2000) and in cucumber, melatonin treatment even provoked the production of ABA during the first 4 days during chilling-stress (Zhao et al. 2017). During drought conditions in apple leaves the pre-treatment of melatonin reduced the ABA amount via the regulation of ABA biosynthesis (Lee et al. 2015).

Under cold stress the exogenous treatment of melatonin dramatically increased the chlorophyll content of the primed plants. The positive effects of melatonin on rubisco activity and photosynthetic electron transport could exert by mitigating the cold stress-induced photosynthetic constraints in barley plants (Li et al. 2016). The reduced ABA content, retarding the senescence process was observed in heat-induced perennial ryegrass leaves after exogenous melatonin treatment (Zhang et al. 2017). In two genotypes of Chinese wild ryegrass (*Elymus nutans* Griseb.), i.e., the cold-tolerant Damxung and the cold-sensitive Gannan, the role of melatonin in ABA-dependent as well as ABA-independent pathways was observed (Fu et al. 2017). Enhanced ABA synthesis was recorded in melatonin-treated plants, and melatonin along with ABA alleviates cold induced oxidative damage, reduce reactive oxygen species (ROS) and malondialdehyde (MDA) levels and increase antioxidant metabolites like ascorbate (AsA) and glutathione (GSH) and other related-enzymes such as superoxide dismutase (SOD), ascorbate peroxidase (APX), catalase (CAT) and glutathione reductase (GR). Melatonin also upregulates the expression of major cold-responsive genes such as CBF and COR genes. ABA and ABA biosynthesis inhibitor fluridone both cannot change endogenous melatonin levels, indicating that ABA might be act as a downstream melatonin signals under the influence of stress response. Thus, cold-induced melatonin can enhance ABA levels via an ABA-dependent pathway.

4.2.5 Ethylene and Melatonin

Melatonin found to be exerts their control over a variety of ripening parameters, including RIN, NOR, CNR, and AP2a by triggering the ethylene production. In tomatoes the carotenoid biosynthesis genes are turned on by melatonin treatment. The transcription of ACO and ACS genes along with transducing elements (EIL1, EIL3, and ERF2) was also enhanced by the melatonin treatment. A differential proteomic test in tomato unveils that melatonin has a remarkable influence on different proteins involved in the ripening pathway (Sun et al. 2016). The treatment of tomatoes with 50 μM of melatonin enriches the fruits production with considerable modifications in their fruit ripening properties, such as fruit softening, flavor, ethylene signaling, lycopene levels, and biosynthesis enzymes than those of untreated tomatoes (Sun et al. 2015). The application of melatonin extends fruit shelf life and enhanced various bioactive substances such as flavonoid, antioxidants, and phenolic acids and also improved fruit quality indicators (Zahedi et al. 2020). In grapevines, melatonin observed to enhance the salt resistance of vines and elevate the ripening of berry, related to ethylene production (Xu et al. 2018). Particularly, melatonin treatment retarded ethylene production in banana leaves, however, the synergistic effect of melatonin and *Fusarium* wilt promote ethylene production (Wei et al. 2017). Therefore, it has been concluded that melatonin regulate ethylene biosynthesis indirectly, and its effect on ethylene production may also be influenced by other factors, possibly through the subsidiary signal molecules.

Xu et al. (2019) studied that the repression of MYB108A expression partly restricted the consequence of melatonin on the stimulation of ethylene synthesis and inhibit melatonin-induced salt tolerance. Moreover, ethylene biosynthesis as well as salt tolerance was also promoted by melatonin through the regulation of ACS1 via MYB108A. In contrast, the inhibitory effect of melatonin on ethylene biosynthesis was also observed in etiolated lupin seedlings (Arnao and Hernández-Ruiz 2007). Melatonin-treated seeds in tomato also produced plants with improved yields and higher lycopene, ascorbic acid, and Ca levels, improved ripening and flavors, whereas the amount of Cu, Fe, N, Mg, Zn, and Mn decreased (Liu et al. 2016). Enhanced citric acid level was also observed in melatonin-treated plants. In melatonin-treated strawberries, post-harvest decay indicators have found to improve along with increased ATP contents, polyphenol levels, and antioxidant activity (Aghdam and Fard 2017).

4.2.6 *Jasmonic Acid and Melatonin*

Melatonin promotes the reduction in jasmonic acid (JA) anabolism and its level in *Brassica napus* under the influence of salinity stress. Moreover, it activates the production of JAZ proteins, a suppressor in the JA signaling pathway, which in turn inhibits the JA mediated response. Consequently, there is a reduced JA response, which increases salt tolerance (Tan et al. 2019). Although melatonin effect on JA levels can fluctuate and observed to lowers or reverses the response of JA through the production of JAZ protein. During a novel study in *Arabidopsis* primary root development found to decrease by high melatonin doses (0.1 and 1.0 mM) (Yang et al. 2021). Jasmonic acid and melatonin are found to involved in responses to cold resistance of plants (Tiwari et al. 2020). The synergistic action of JA and melatonin has been reported during cold resistance in tomato plants. It has been observed that JA and melatonin both cumulate under the influence of cold stress while foliar applications of methyl jasmonates (MeJA) and melatonin in tomato plants found to enhance cold tolerance by increasing Fv/Fm, decreasing relative electrolyte leakage (EL), as well as reducing H_2O_2 accumulation. The inhibition of melatonin biosynthesis reduced MeJA-dependent cold tolerance, whereas melatonin accumulation has been decreased by the inhibition of JA biosynthesis in tomato plants under cold stress. Interestingly, exogenous melatonin promoted accumulation of JA, while inhibition of melatonin biosynthesis dramatically reduced the accumulation of JA in tomato plants during the cold stress. These findings suggest that JA and melatonin act synergistically during cold tolerance forming a positive feedback loop, promoting the cold responses in tomato plants. Moreover, MeJA induced the expressions of two melatonin biosynthesis genes, i.e., SlSNAT and SlAMST, while the repression of SlMYC2, a potential JA signaling regulator, suppressed the expressions of SlSNAT and SlAMST under cold stress. Furthermore, the inhibition of SlMYC2 declined MT accumulation, reduced Fv/Fm, and increased EL in cold-stressed tomato plants (Ding et al. 2022).

4.2.7 Salicylic Acid and Melatonin

Melatonin and salicylic acid (SA) have been observed to play various roles in plant physiological processes and stress responses. SA regulated different biotic stress responses in plants (Khan et al. 2021). The implicated antioxidative metabolism during plant stress response regulated by melatonin, along with NO exhibiting a crucial role, as well as SA also implicated through a downstream signaling pathway. Evidently melatonin is the crucial player which directly or indirectly enhances SA, JA, and ethylene, thereby activating the pathogen defense responses. An increased in melatonin and SA level was observed in *Arabidopsis* plant infected by *Pseudomonas syringae* DC3000. However, the melatonin and SA levels were reduced in SNAT knockout mutants and which were also more susceptible to pathogen attacks (Lee et al. 2015). The exogenously administered melatonin promotes the synthesis of SA and NO, thereby defending plants from the infection of tobacco mosaic virus (Zhao et al. 2019). Conclusively melatonin and salicylic acid both are found to involved in responses to abiotic stress situations, including a significant enhancement in the water level during drought stress, improved biosynthesis of photosynthetic pigments and photosynthetic rate, an increase in metabolites and antioxidant enzymes in order to balance the redox status, osmotic balance to reduce membrane injury under stress conditions, as well as in growth promotion, increased productivity and yield (Wang et al. 2018; Al Azzawi et al. 2020).

4.3 Conclusion

In this chapter we have summarized the possible role of melatonin under normal physiological conditions and during biotic and abiotic stresses. Melatonin crosstalk with other phytohormones is also described briefly in this chapter. Melatonin can significantly enhance the crops quality and quantity because they alter the morphology, physiology and eventually enhancing the biomass at specific concentration. Therefore, it can be useful in the overproduction of agricultural biomass at global scale during changing environmental conditions.

References

Aghdam MS, Fard JR (2017) Melatonin treatment attenuates postharvest decay and maintains nutritional quality of strawberry fruits (Fragaria anannasa cv. Selva) by enhancing GABA shunt activity. Food Chem 221:1650–1657

Al Azzawi TNI, Khan M, Hussain A, Shahid M, Imran QM, Mun BG, Lee S, Yun BW (2020) Evaluation of Iraqi rice cultivars for their tolerance to drought stress. Agronomy 10:1782

Arnao MB, Hernández-Ruiz J (2007) Melatonin promotes adventitious and lateral root regeneration in etiolated hypocotyls of *Lupinus albus* L. J Pineal Res 42:147–152

Arnao MB, Hernández-Ruiz J (2009) Protective effect of melatonin against chlorophyll degradation during the senescence of barley leaves. J Pineal Res 46:58–63

Arnao MB, Hernández-Ruiz J (2017) Growth activity, rooting capacity, and tropism: three auxinic precepts fulfilled by melatonin. Acta Physiol Plant 39:127

Arnao MB, Hernández-Ruiz J (2018) Melatonin and its relationship to plant hormones. Ann Bot 121(195):207

Arnao MB, Hernández-Ruiz J (2021) Melatonin as a regulatory hub of plant hormone levels and action in stress situations. Plant Biol 23:7–19

Bajwa VS, Shukla MR, Sherif SM, Murch SJ, Saxena PK (2014) Role of melatonin in alleviating cold stress in *Arabidopsis thaliana*. J Pineal Res 56:238–245

Ding F, Ren L, Xie F, Wang M, Zhang S (2022) Jasmonate and melatonin act synergistically to potentiate cold tolerance in tomato plants. Front Plant Sci 12:763284. PMID: 35069620; PMCID: PMC8776829. https://doi.org/10.3389/fpls.2021.763284

Fan J, Xie Y, Zhang Z, Chen L (2018) Melatonin: a multifunctional factor in plants. Int J Mol Sci 19:1528

Fernández-Marcos M, Sanz L, Lewis DR, Muday GK, Lorenzo O (2011) Nitric oxide causes root apical meristem defects and growth inhibition while reducing PIN-FORMED 1 (PIN1)-dependent acropetal auxin transport. Proc Natl Acad Sci U S A 108:18506–18511

Fu J, Wu Y, Miao Y et al (2017) Improved cold tolerance in *Elymus nutans* by exogenous application of melatonin may involve ABA-dependent and ABA-independent pathways. Sci Rep 7:39865

Hwang OJ, Back K (2022) Exogenous gibberellin treatment enhances melatonin synthesis for melatonin-enriched rice production. Biomol Ther 12(2):198. https://doi.org/10.3390/biom12020198

Jia W, Zhang J (2000) Water stress-induced abscisic acid accumulation in relation to reducing agents and sulfhydryl modifiers in maize plant. Plant Cell Environ 23:1389–1395

Khan M, Nazar T, Pande A, Mun BG, Lee D, Hussain A, Lee BH, Yun BWJF (2021) The role of nitric oxide-induced ATILL6 in growth and disease resistance in Arabidopsis thaliana. Front Plant Sci 12:1314

Kim M, Seo H, Park C, Park WJ (2016) Examination of the auxin hypothesis of phytomelatonin action in classical auxin assay systems in maize. J Plant Physiol 190:67–71

Koyama FC, Carvalho TLG, Alves E et al (2013) The structurally related auxin and melatonin tryptophan-derivatives and their roles in *Arabidopsis thaliana* and in the human malaria parasite plasmodium falciparum. J Eukaryot Microbiol 60:646–651

Lee HY, Byeon Y, Tan DX, Reiter RJ, Back K (2015) *Arabidopsis serotonin* N-acetyltransferase knockout mutant plants exhibit decreased melatonin and salicylic acid levels resulting in susceptibility to an avirulent pathogen. J Pineal Res 58:291–299

Li C, Liang B, Chang C, Wei Z, Zhou S, Ma F (2016) Exogenous melatonin improved potassium content in *malus* under different stress conditions. J Pineal Res 61:218–229

Liang C, Li A, Yu H, Li W, Liang C, Guo S et al (2017) Melatonin regulates root architecture by modulating auxin response in rice. Front Plant Sci 8:134

Liu J, Zhang R, Sun Y, Liu Z, Jin W, Sun Y (2016) The beneficial effects of exogenous melatonin on tomato fruit properties. Sci Hortic 207:14–20

Liu S, Huang H, Huber DJ, Pan Y, Shi X, Zhang Z (2020) Delay of ripening and softening in 'Guifei' mango fruit by postharvest application of melatonin. Postharvest Biol Technol 163: 111136

Ma X, Zhang J, Burgess P, Rossi S, Huang B (2018) Interactive effects of melatonin and Cytokinin on alleviating drought-induced leaf senescence in creeping bent grass (*Agrostis stolonifera*). Environ Exp Bot 145:1–11

Mao J, Niu C, Li K, Chen S, Tahir MM, Han M, Zhang D (2020) Melatonin promotes adventitious root formation in apple by promoting the function of MdWOX11. BMC Plant Biol 20:536

Mukherjee S, David A, Yadav S, Baluska F, Bhatla SC (2014) Salt stress induced seedling growth inhibition coincides with differential distribution of serotonin and melatonin in sunflower seedling roots and cotyledons. Physiol Plant 152:714–728

Nawaz MA, Huang Y, Bie Z et al (2016) Melatonin: current status and future perspectives in plant science. Front Plant Sci 6:1230

Park WJ (2011) Melatonin as an endogenous plant regulatory signal: debates and perspectives. J Plant Biol 54:143–149

Posmyk MM, Bałabusta M, Wieczorek M, Sliwinska E, Janas KM (2009) Melatonin applied to cucumber (Cucumis sativus L.) seeds improves germination during chilling stress. J Pineal Res 46:214–223

Reiter RJ, Tan DX, Galano A (2014) Melatonin: exceeding expectations. Physiology (Bethesda) 56:371–381

Ren S, Rutto L, Katuuramu D (2019) Melatonin acts synergistically with auxin to promote lateral root development through fine tuning auxin transport in *Arabidopsis thaliana*. PloS One 14(8): e0221687. PMID: 31461482; PMCID: PMC6713329. https://doi.org/10.1371/journal.pone.0221687

Sanz L, Albertos P, Mateos I, Sánchez-Vicente I, Lechón T, Fernández-Marcos M, Lorenzo O (2015) Nitric oxide (NO) and phytohormones crosstalk during early plant development. J Exp Bot 66:2857–2868

Sharma A, Zheng B (2019) Melatonin mediated regulation of drought stress: physiological and molecular aspects. Plan Theory 8:190

Shi H, Jiang C, Ye T et al (2015a) Comparative physiological, metabolomic, and transcriptomic analyses reveal mechanisms of improved abiotic stress resistance in bermudagrass [*Cynodon dactylon* (L). Pers.] by exogenous melatonin. J Exp Bot 66:681–694

Shi HT, Reiter RJ, Tan DX, Chan ZL (2015b) Indole-3-acetic acid inducible 17 positively modulates natural leaf senescence through melatonin-mediated pathway in *Arabidopsis*. J Pineal Res 58:26–33. https://doi.org/10.1111/jpi.12188

Shi H, Tan DX, Reiter RJ, Ye T, Yang F, Chan Z (2015c) Melatonin induces class A1 heat-shock factors (HSFA 1s) and their possible involvement of thermo tolerance in *Arabidopsis*. J Pineal Res 58:335–342

Sun C, Lu L, Liu L, Liu W, Yu Y, Liu X, Hu Y, Jin C, Lin X (2014) Nitrate reductase-mediated early nitric oxide burst alleviates oxidative damage induced by aluminum through enhancement of antioxidant defenses in roots of wheat (*Triticum aestivum*). New Phytol 201:1240–1250
Sun QQ, Zhang N, Wang J et al (2015) Melatonin promotes ripening and improves quality of tomato fruit during postharvest life. J Exp Bot 66:657–668
Sun Q, Zhang N, Wang J et al (2016) A label-free differential proteomics analysis reveals the effect of melatonin on promoting fruit ripening and anthocyanin accumulation upon postharvest in tomato. J Pineal Res 61:138–153
Tan X, Long W, Zeng L, Ding X, Cheng Y, Zhang X, Zou X (2019) Melatonin-induced transcriptome variation of rapeseed seedlings under salt stress. Int J Mol Sci 20:5355
Tijero V, Munoz P, Munné-Bosch S (2019) Melatonin as an inhibitor of sweet cherries ripening in orchard trees. Plant Physiol Biochem 140:88–95
Tiwari RK, Lal MK, Naga CK, Kumar R, Chourasia KN, Subhash S et al (2020) Emerging roles of melatonin in mitigating abiotic and biotic stress of horticultural crops. Sci Hortic 272:109592. https://doi.org/10.1016/j.scienta.2020.109592
Wang Y, Reiter RJ, Chan Z (2018) Phytomelatonin: a universal abiotic stress regulator. J Exp Bot 69:963–974
Wang Y, Li J, Yang L et al (2022) Melatonin antagonizes cytokinin response to stimulate root growth in *Arabidopsis*. J Plant Growth Regul 42:1833. https://doi.org/10.1007/s00344-022-10663-9
Weeda S, Zhang N, Zhao X et al (2014) *Arabidopsis* transcriptome analysis reveals key roles of melatonin in plant defense systems. PloS One 9:e93462
Wei YX et al (2017) Identification, transcriptional and functional analysis of heat-shock protein 90s in banana (*Musa acuminata* L.) highlights their novel role in melatonin-mediated plant response to *Fusarium* wilt. J Pineal Res 62:1–12. https://doi.org/10.1111/jpi.12367
Wen D, Gong B, Sun S et al (2016) Promoting roles of melatonin in adventitious root development of *Solanum lycopersicum* L. by regulating auxin and nitric oxide signaling. Front Plant Sci 7: 718
Xu L, Yue Q, Xiang G, Bian F, Yao Y (2018) Melatonin promotes ripening of grape berry via increasing the levels of ABA, H_2O_2, and particularly ethylene. Hortic Res 5:41–51. https://doi.org/10.1038/s41438-018-0045-y
Xu L, Xiang G, Sun Q, Ni Y, Jin Z, Gao S, Yao Y (2019) Melatonin enhances salt tolerance by promoting MYB108A-mediated ethylene biosynthesis in grapevines. Hortic Res 6:114. PMID: 31645968; PMCID: PMC6804660. https://doi.org/10.1038/s41438-019-0197-4
Yang L, Sun Q, Wang Y, Chan Z (2021) Global transcriptomic network of melatonin regulated root growth in *Arabidopsis*. Gene 764:145082. https://doi.org/10.1016/j.gene.2020.145082
Yu M, Lamattina L, Spoel SH, Loake GJ (2014) Nitric oxide function in plant biology: a redox cue in deconvolution. New Phytol 202:1142–1156
Zahedi SM, Hosseini MS, Abadía J, Marjani M (2020) Melatonin foliar sprays elicit salinity stress tolerance and enhance fruit yield and quality in strawberry (*Fragaria ananassa* Duch.). Plant Physiol Biochem 149:313–323
Zhang N, Zhang HJ, Zhao B, Sun QQ, Cao YY, Li R, Wu XX, Weeda S, Li L, Ren S (2014) The RNA-seq approach to discriminate gene expression profiles in response to melatonin on cucumber lateral root formation. J Pineal Res 56:39–50
Zhang J, Shi Y, Zhang X, Du H, Xu B, Huang B (2017) Melatonin suppression of heat-induced leaf senescence involves changes in abscisic acid and Cytokinin biosynthesis and signaling pathways in perennial ryegrass (*Lolium perenne* L.). Environ Exp Bot 138:36–45
Zhao H, Zhang K, Zhou X, Xi L, Wang Y, Xu H, Pan T, Zou Z (2017) Melatonin alleviates chilling stress in cucumber seedlings by up-regulation of CsZat12 and modulation of polyamine and abscisic acid metabolism. Sci Rep 7:4998

Zhao L, Chen L, Gu P, Zhan X, Zhang Y, Hou C, Wu Z, Wu YF, Wang QC (2019) Exogenous application of melatonin improves plant resistance to virus infection. Plant Pathol 68:1287–1295

Zuo B, Zheng X, He P et al (2014) Over expression of MzASMT improves melatonin production and enhances drought tolerance in transgenic *Arabidopsis thaliana* plants. J Pineal Res 57:408–417

Chapter 5
Interaction Between Melatonin and Other Signaling Molecules in Plants

Yanliang Guo and Hao Li

Abstract Numerous studies have proven evidence that melatonin plays an important regulatory role in plant growth, development, and defense against various biotic and abiotic stresses. As well as melatonin, plant hormones and many second messengers, such as calcium (Ca^{2+}), reactive oxygen species (ROS), nitric oxide (NO), and hydrogen sulfide (H_2S), also play important roles in the regulation of various physiological processes in plants. In recent years, increasing studies have indicated that melatonin interacts with plant hormones and many second messengers to regulate multiple physiological processes. The role of melatonin in regulating seed germination involves abscisic acid (ABA), gibberellins (GA), Ca^{2+} signal, and H_2O_2; in regulating stomatal movement involves ABA, Ca^{2+}, ROS, and H_2S; in regulating rhizogenesis involves auxin, ROS, and NO; in regulating fruit ripening involves ethylene (ETH), ROS, and NO; in regulating plant senescence involves ABA, cytokinins (CKs), Ca^{2+}, ROS, and NO; in regulating tolerance to abiotic stress involves CK, ABA, ETH, jasmonic acid (JA), and all secondary signals mentioned above; in regulating disease resistance involves ETH, JA, salicylic acid (SA), ROS, and NO. Plant hormones and the second messengers also interact with each other, forming complex regulatory networks, to regulate multiple physiological processes in plants.

Keywords Calcium signal · Hydrogen sulfide · Melatonin · Nitric oxide · Plant hormones · Reactive oxygen species

---

Y. Guo · H. Li (✉)
State Key Laboratory of Crop Stress Biology for Arid Areas, College of Horticulture, Northwest A&F University, Yangling, Shaanxi, People's Republic of China
e-mail: lhyy@nwafu.edu.cn

A. Sharma, G. J. Ahammed (eds.), *Melatonin in Plants: Role in Plant Growth, Development, and Stress Response*, Plant Life and Environment Dynamics,
https://doi.org/10.1007/978-981-99-8051-2_5

5.1 Introduction

Melatonin is a universal and pleiotropic molecule with many diverse actions in plants (Altaf et al. 2021; Arnao and Hernandez-Ruiz 2014; Fan et al. 2018). The identification of melatonin receptor (CAND2/PMTR1) in *Arabidopsis* provides a piece of strong evidence for considering melatonin as a new plant hormone (Arnao and Hernández-Ruiz 2018a; Wei et al. 2018). Like melatonin, plant hormones play crucial roles in regulating plant growth, development, and adaptation to adverse environmental conditions. An increasing number of studies have shown that melatonin and plant hormones cooperatively or antagonistically regulate various physiological processes in plants (Arnao and Hernández-Ruiz 2018b). For instance, melatonin interacts with auxin to regulate lateral and adventitious root formation and growth; interacts with cytokinins (CKs) to regulate senescence and tolerance against various abiotic stresses; interacts with abscisic acid (ABA) to regulate seed germination, stomatal movement, senescence, and abiotic stress tolerance; interacts with gibberellins (GAs) to regulate seed germination; and interacts with ethylene (ETH) to regulate fruit ripening, postharvest processes, and resistance to abiotic stresses and diseases. In plant response to the pathogen, melatonin functions as a critical regulator to induce an immune response, together with jasmonic acid (JA), salicylic acid (SA), and ETH.

As well as plant hormones, many second messengers, such as calcium (Ca^{2+}), reactive oxygen species (ROS), nitric oxide (NO), and hydrogen sulfide (H_2S), also play important roles in the regulation of multiple physiological processes such as seed germination, stomatal movement, delay of leaf senescence, and stress defense. These second messengers also are involved in the melatonin-mediated regulation of various physiological processes in plants.

Ca^{2+} is one of the most important second messengers that can regulate plant growth, development, and response to various environmental stresses (Berridge et al. 2000). Cytosolic-free Ca^{2+} ($[Ca^{2+}]_{cyt}$) homeostasis is maintained by Ca^{2+}-permeable channels and membrane transporters that mediate Ca^{2+} influx into and efflux from plant cells, respectively (Dodd et al. 2010). Ca^{2+} signal functions by eliciting transient $[Ca^{2+}]_{cyt}$ changes through Ca^{2+} channels and, or transporters and is subsequently decoded by multiple Ca^{2+}-binding sensors, such as calmodulins (CaMs), calcineurin B-likes (CBLs), CaM-like proteins (CMLs), Ca^{2+}/CaM-dependent protein kinases (CCaMKs), and Ca^{2+}-dependent protein kinases (CDPKs) (Dodd et al. 2010). It has been demonstrated that the Ca^{2+} signal is involved in the melatonin-mediated regulation of multiple physiological processes such as seed germination, stomatal movement, plant senescence, and stress responses in plants.

Reactive oxygen species (ROS) are generated in multiple organelles in plant cells, especially when physiological processes, such as photosynthesis, respiration, and photorespiration, are impaired under adverse environmental conditions (Dat et al. 2000). As strong oxidizers, excessive ROS can cause oxidative stress that further impair plant cells and various physiological processes (Bose et al. 2014). Melatonin acts as an effective antioxidant to directly remove some ROS or modulate the antioxidant system including antioxidant enzymes and antioxidants, and consequently alleviates oxidative stress (Zhang and Zhang 2014; Sharma and Zheng 2019; Sharma et al. 2020). Excessive ROS are harmful; however, moderate levels of ROS, especially hydrogen peroxide (H_2O_2) at the apoplast generated by *RBOH* encoded NADPH oxidase, act as critical second messengers and play important roles in regulating plant growth, development, and defending against stresses (Baxter et al. 2014). For instance, *RBOH1*-dependent H_2O_2 production induces the accumulation of brassinazole-resistant 1 (BZR1), a central component of brassinosteroids signaling, leading to increases in the levels of CBF transcripts and subsequent cold tolerance in *Solanum lycopersicum* (Fang et al. 2021). ROS, especially H_2O_2, play important roles in melatonin regulating seed germination, stomatal movement, rhizogenesis, fruit ripening, plant senescence, and plant adaption to adverse environments.

Nitric oxide (NO) is an imperative gas signaling molecule that plays a critical role in regulating a variety of physiological and biochemical processes in plants, such as seed germination, stomatal movement, flowering, leaf senescence, and environmental stress response (Mur et al. 2012). NO in plants is mainly produced through the NO synthase (NOS) and nitrate reductase (NR) pathways that with L-arginine and nitrite as a substrate, respectively (Lozano-Juste and León 2010). NO functions through multiple processes, such as regulation of target protein activities through S- and/or metal-nitrosylation and induction of transient $[Ca^{2+}]_{cyt}$ rises by activating Ca^{2+} influx channels (Besson-Bard et al. 2008). The pathways that switch off NO signal includes S-nitrosoglutathione reductase (GSNOR) and inhibition of the channels (e.g., cyclic nucleotide-gated channel 2) that mediated Ca^{2+} influx. It has been reported that NO is involved in melatonin-mediated regulation of rhizogenesis, fruit ripening, plant senescence, and stress responses.

Hydrogen sulfide (H_2S), a small lipophilic gaseous molecule, is considered the third most important gas messenger after carbon monoxide (CO) and NO in plants (Fotopoulos et al. 2015; Guo et al. 2016). Like NO and H_2O_2, H_2S at high concentrations has toxicity leading to inhibition of cytochrome oxidase but it at low concentrations can serve as a signal molecule (Li 2015). H_2S biosynthesis involves four enzymes including L-cysteine desulfhydrase (L-CDes), D-CDes, sulfite reductase, and β-cyanoalanine synthase, while H_2S degradation is mediated by cysteine synthase (Li 2015). H_2S has a major role in regulating seed germination, lateral root formation, stomatal movement, photosynthesis, maturation, plant senescence, and adaption to various adverse environmental conditions such as drought, extreme temperature, salt, flood, and heavy metal stress (Pandey and Gautam 2020; Xuan et al. 2020). It also induces cross-adaptation through interacting with other signaling molecules such as Ca^{2+}, NO, methylglyoxal, and H_2O_2 and modulating the expression of stress defense-related genes such as *mitogen-activated protein kinases* (*MAPKs*). The crosstalk between melatonin and H_2S has been explored in the regulation of stomatal movement and tolerance to various abiotic stresses.

5.2 Melatonin and Plant Hormones

5.2.1 Melatonin and Auxin

Melatonin and indole-3-acetic acid (IAA), the main natural auxin, share the same biosynthetic precursor, tryptophan, have similar structural moieties, and thus have some similar functions. Many studies have identified the auxin-like function of melatonin which can induce the formation of both lateral and adventitious roots and the growth of roots and shoots in various plant species (Arnao and Hernández-Ruiz 2017). Generally, the application of exogenous melatonin can increase the levels of endogenous IAA (Wen et al. 2016). However, melatonin is unable to regulate the expression of auxin-inducible genes in auxin-inducible marker DR5: GUS transgenic *Arabidopsis*, suggesting that melatonin and auxin seem to function in parallel in inducing the formation of lateral and adventitious roots (Arnao and Hernandez-Ruiz 2014). In *Arabidopsis*, melatonin downregulated the expression of *AUXIN1/LIKE-AUX1* (*AUX1/LAX*) that encodes auxin influx carriers. AUX1 regulates lateral root initiation and gravitropism, root hair development, and leaf phyllotaxy, suggesting that melatonin can alter the expression of genes encoding auxin influx carriers that drive auxin homeostasis and consequently regulate auxin action in some developmental programs (Arnao and Hernández Ruiz 2015; Wang et al. 2016). In rice, exogenous melatonin is able to alter the expression of many auxin-related transcription factors and it has been confirmed that melatonin plays an important role in shaping root architecture through modulating auxin response (Liang et al. 2017). Serotonin *N*-acetyltransferase (SNAT) and *N*-acetylserotonin *O*-methyltransferase (ASMT) are two critical melatonin biosynthetic enzymes. Contrary to the effects of exogenous melatonin, overexpression of apple *MzASMT1* in *Arabidopsis* and ovine *SNAT/ASMT* in tomato increased melatonin production but reduced the levels of IAA (Zuo et al. 2014; Arnao and Hernández-Ruiz 2018a). Unsurprisingly, melatonin and IAA share the same precursor and thus increased biosynthesis of endogenous melatonin could inhibit IAA biosynthesis, due to substrate competition. Overexpression of melatonin biosynthetic genes *SNAT* and *ASMT* in the transgenic plants induced auxin-like responses, such as induction of root formation and growth (Zuo et al. 2014; Arnao and Hernández-Ruiz 2018a).

In *Arabidopsis*, the application of melatonin at high concentration (600 μM) inhibited root growth by reducing root meristem size, auxin biosynthesis, and the expression of polar auxin transport-related protein PINFORMED (PIN) (Wang et al. 2016). Application of triiodobenzoic acid that inhibits polar auxin transport did not enhance the melatonin-induced decrease in root meristem size, indicating that high concentration of melatonin-mediated inhibition of root meristem and growth in *Arabidopsis* is partially dependent on the repression of biosynthesis and polar transport of auxin. Overproduction of melatonin also induced reduction of apical dominance but increase in branching, which are contrary to auxin response, suggesting that melatonin and IAA play different roles in the regulation of apical dominance function (Zuo et al. 2014).

5.2.2 Melatonin and CKs

A few studies have reported the positive crosstalk between melatonin and CKs in regulating plant senescence. Application of melatonin at a suitable concentration can delay the dark- and ABA-induced senescence of barley and melon, respectively, inhibiting the degradation of chlorophyll in leaves (Arnao and Hernández-Ruiz 2009; Guo et al. 2023). Like melatonin, kinetin (a synthetic CK) at different concentrations can reduce chlorophyll loss and thus retard the senescence process of leaf sections (Arnao and Hernández-Ruiz 2009). Under normal conditions, melatonin application did not induce changes in CK levels (Zhang et al. 2017b). In contrast, under heat stress, exogenous melatonin promoted growth and delayed leaf senescence, accompanied by increased endogenous CK contents and upregulated expression of CK biosynthetic genes and CK signaling response transcription factors, indicating that the melatonin-induced delay of leaf senescence under heat stress involves the activation of CK biosynthesis and signaling (Zhang et al. 2017b). Similarly, under drought stress, melatonin inhibited leaf senescence in creeping bentgrass and upregulated the expression of genes involved in CK biosynthesis and signaling, suggesting that melatonin-mediated inhibition of leaf senescence caused by drought in creeping bentgrass is correlated with its positive regulation of CK biosynthesis and signaling (Ma et al. 2018).

5.2.3 Melatonin, ABA, and GA

GA and ABA are considered as the two most critical phytohormones that function antagonistically to regulate seed dormancy and germination (Footitt et al. 2011). Melatonin upregulates the expression of GA biosynthetic genes *GA 20-oxidase* (*GA20ox*) and *GA 3-oxidase* (*GA3ox*) and consequently increases GA_3 and GA_4 levels, promoting cucumber seed germination under saline conditions (Zhang et al. 2014a, b). Also, melatonin upregulates the expression of genes (two *CYP707 monooxygenases*) involved in ABA catabolism and downregulates the expression of *9-cis-epoxycarotenoid dioxygenase* (*NCED*), a key gene in ABA biosynthesis, leading to decreased levels of ABA during seed germination under salt stress (Zhang et al. 2014b). Similarly, in apple plants exposed to drought stress, melatonin decreases the levels of endogenous ABA by regulating the genes involved in ABA biosynthesis and catabolism as mentioned above (Li et al. 2015). Recently, Li et al. (2021b) found that melatonin offsets ABA to induce melon seed germination by promoting ABA catabolism and/or GA_3 biosynthesis and thus maintaining ABA and GA_3 balance.

Li et al. (2016b) investigated the interaction between melatonin and ABA in regulating the cold tolerance of drought-primed barley plants. The levels of endogenous melatonin and ABA in wild-type barley leaves were significantly increased by all stressing treatments. Melatonin application increased ABA levels in the drought-primed plants that were exposed to cold. Drought priming increased the melatonin levels in an ABA-deficient barley mutant, indicating that melatonin response to drought priming under cold is ABA-independent. Therefore, melatonin-induced cold tolerance in drought-primed barley plants essentially involves the modulation of ABA levels. In grafted watermelon plants under cold stress, cold tolerant root-stocks enhanced cold tolerance of watermelon scion, accompanied by increased levels of ABA and melatonin. Application of ABA and melatonin increased the levels of endogenous melatonin and ABA, respectively, and enhanced cold tolerance of self-grafted watermelon plants. Inhibition of biosynthesis of ABA and melatonin blocked the cold tolerant rootstocks-induced the accumulation of melatonin and ABA, respectively, as well as cold tolerance. These results suggest that melatonin and ABA interact with each other, forming a positive feedback loop, to enhance cold tolerance of grafted watermelon plants (Guo et al. 2021). In Chinese wild ryegrass response to cold stress, melatonin increased ABA accumulation, whereas, ABA did not induce changes in melatonin accumulation, suggesting that ABA functions as a downstream mediator of melatonin (Skinner et al. 2005). In Chinese wild oats, melatonin enhanced cold tolerance through both ABA-dependent and ABA-independent pathways (Fu et al. 2017). In perennial ryegrass response to heat stress, melatonin alleviated heat-induced leaf senescence through downregulating the expression of ABA biosynthetic and signaling genes and reducing endogenous ABA levels (Zhang et al. 2017a). Therefore, melatonin counteracts ABA in heat-induced senescence through inhibiting ABA biosynthesis and its signaling pathway factors. More recently, Guo et al. (2023) have demonstrated that melatonin can alleviate ABA-induced senescence of melon leaves via Ca^{2+}-mediated K^+ efflux in H_2O_2-dependent way.

5.2.4 Melatonin and ETH

Studies on melatonin and ETH crosstalk in plants mainly focus on the regulation of fruit ripening and stress response. As well as ETH, melatonin application promoted postharvest ripening and improved quality of tomato fruit, and such role of melatonin is associated with its upregulation of genes involved in cell wall degradation and the biosynthesis of lycopene and compounds related to aroma and flavor (Sun et al. 2015, 2016). Moreover, melatonin promoted ETH generation by inducing the expression of *1-aminocycopropane-1-carboxylic acid (ACC) synthase*, a key gene involved in ETH biosynthesis, indicating that melatonin-induced promotion of postharvest fruit ripening is associated with ETH generation and signaling. Furthermore, melatonin decreased the DNA methylation levels of CpG island and expression of *Constitutive triple response 1* (*CTR1*) but induced that of *1-aminocyclopropane-1-carboxylate synthase 10* (*ACS10*) and *ethylene responsive factor A1* (*SlERF-A1*) in tomato fruits, and thus might activate ETH signaling and promote ripening (Shan et al. 2022). Therefore, melatonin-mediated regulation of DNA methylation of CpG islands of ETH biosynthesis- and signaling-related genes may contribute to tomato fruit ripening. However, melatonin as well as auxin showed an inhibitory effect on ACC oxidase activity and subsequent ETH biosynthesis in etiolated lupin (Arnao and Hernández-Ruiz 2007; Kim et al. 2001).

In grapevines, Xu et al. (2019) found that the application of exogenous melatonin and ethylene precursor ACC enhanced tolerance to salt stress. Melatonin-induced ACC accumulation and ETH generation and upregulated the expression of *MYB108A* and *ACS1*. *MYB108A* activated the transcription of *ACS1* by directly binding to its promoter and thus induced ACC biosynthesis and ETH generation. The suppression of *MYB108A* expression inhibited melatonin-induced ETH production and salt tolerance, suggesting that melatonin positively regulates ETH production and salt tolerance through the activation of *MYB108A*-dependent *ACS1*. In addition, melatonin upregulated the expression of genes related to pathogen defense and ETH signaling and induced innate immune responses to the infection of bacterial pathogen *Pseudomonas syringae* (*Pst* DC3000) in *Arabidopsis* and tobacco. The positive effects of melatonin on pathogen defense were abolished or attenuated in *Arabidopsis* mutants defective in ETH signaling, suggesting that ETH is involved in melatonin-induced pathogen resistance (Lee et al. 2014).

5.2.5 Melatonin, JA and SA

Numerous studies have demonstrated that melatonin plays a crucial role in plant immune response and this function of melatonin is related to JA, and SA, as well as ETH (Arnao and Hernández-Ruiz 2018b). Melatonin treatment of apple trees could reduce the number of diseased leaves, increase chlorophyll content, greatly resist fungal infection, reduce lesions, inhibit the spread of pathogens, and reduce losses (Yin et al. 2013). In terms of plant disease resistance regulation, as well as ETH, JA and SA are two of the most extensively studied plant hormones. JA/ET-dependent plant defensin PDF1.2 and pathogenesis-related protein 1 (PR1) are well known markers that reflect JA/ET and SA signaling, respectively, in *Arabidopsis*. Melatonin induced the expression of pathogen defense-related genes, such as *PDF1.2* and *PR1*, and defense responses to *Pst* DC3000 in *Arabidopsis* and tobacco (Lee et al. 2014, 2015). Knockout of *SNAT* not only reduced melatonin levels, but also lower SA levels, along with lower resistance to the pathogen. Moreover, the positive effects of melatonin on pathogen defense were abolished or attenuated in *Arabidopsis* mutants defective in SA signaling, suggesting that SA is required for melatonin-induced defense response to *Pst* DC3000 attack. However, Liu et al. (2021) reported that melatonin reduced the accumulation of JA and conjugate JA-isoleucine in response to *Pst* DC3000 in wild-type *Arabidopsis* leaves and enhanced *Pst* DC3000 resistance of JA- and ETH-insensitive mutants. The resistance of JA- and ETH-insensitive mutants to *Pst* DC3000 implies that JA/ET signaling antagonizes SA-mediated disease resistance (Kunkel and Brooks 2002). Therefore, melatonin seems to regulate plant immunity against pathogens via an SA/JA/ETH-dependent pathway, although the crosstalk between melatonin and JA lacks direct evidence (Arnao and

Hernández-Ruiz 2018b). In addition, melatonin induced the expression of defense-related genes and defense response to gray mold in cherry tomato fruit during storage, accompanied by increased SA biosynthetic enzyme activities and subsequent SA accumulation (Li et al. 2022). The positive effects of melatonin on pathogen defense were completely abolished by an inhibitor of SA biosynthesis, suggesting that SA plays an important role in melatonin inducing protection against gray mold.

The interaction between melatonin and JA also plays a critical role in regulating plant adaption to various abiotic stresses. Cold tolerant rootstock induced watermelon shoot tolerance against cold, accompanied by increased levels of melatonin and methyl jasmonate (MeJA), a JA derivative (Li et al. 2021a). Application of both melatonin and MeJA enhanced cold tolerance. Melatonin promoted the accumulation of MeJA, which in turn induced melatonin biosynthesis, forming a positive feedback loop in response to cold. These results suggest that melatonin and MeJA interaction plays a critical role in grafting-induced cold tolerance. Similarly, in tomato plants, the accumulation of JA and melatonin were induced in response to cold stress (Ding et al. 2022). Application of melatonin or MeJA on leaves induced cold tolerance. Inhibition of JA biosynthesis inhibited melatonin accumulation while inhibition of melatonin biosynthesis inhibited MeJA-induced tolerance in tomato plants under cold conditions, suggesting that melatonin and JA act synergistically to enhance tomato tolerance against the cold.

5.3 Melatonin and Second Messengers

5.3.1 Melatonin and Ca^{2+} Signal

An increasing number of studies have shown that $[Ca^{2+}]_{cyt}$ transient serves as a crucial signal in melatonin-mediated physiological processes, such as stomatal closure, seed germination, postharvest physiological deterioration, and defense responses. Wei et al. (2018) found that melatonin induces Ca^{2+} inward flow through its receptor CAND2/PMTR1 in *Arabidopsis*, thereby inducing stomatal closure. Li et al. (2021b) found that Ca^{2+} signaling is involved in the melatonin-induced promotion of seed germination under ABA stress by inducing GA_3 biosynthesis and ABA catabolism. They also discovered that melatonin counteracts the effects of ABA to delay leaf senescence via triggering $[Ca^{2+}]_{cyt}$ accumulation (Guo et al. 2023).

In regulating plant response to environmental stresses, Imran et al. (2022) showed that melatonin enhances plant resistance to heavy metal toxicity by regulating the expression of *CDPK* to promote plant cell viability and Ca^{2+} signaling. In response to salt stress, melatonin can induce Ca^{2+} burst and upregulate the expression of genes related to Ca^{2+} signal transduction (Tian et al. 2022). Vafadar et al. (2020a, 2021) found that melatonin and Ca^{2+} may function as long-distance signals to induce systemic tolerance against salt stress and Ca^{2+}/CaM signaling mediates melatonin-induced photosynthesis and antioxidant capacity under salinity stress in *Dracocephalum kotschyi*. They also revealed that Ca^{2+} influx into the cells and subsequent Ca^{2+}/CaM signaling is required for melatonin-mediated induction of phenolic production (Vafadar et al. 2020b). In response to cold stress, melatonin can alter the expression of Ca^{2+} signaling-related genes, such as *CIPK16*, *CML23,* and *CML11*, in *Arabidopsis thaliana* and upregulate the expression of *CDPK25* and *Calmodulin-binding protein* in watermelon (Li et al. 2017; Weeda et al. 2014). Ma et al. (2022) found that melatonin-induced growth and photosynthetic capacity rely on Ca^{2+} signaling in cucumber seedlings under cold stress. In addition, an interactive positive feedback regulatory relationship between H_2O_2 and Ca^{2+} mediated the melatonin-induced CBF pathway and subsequent cold resistance in watermelon (Chang et al. 2021). However, there are few studies on the interaction between melatonin and Ca^{2+} signal in regulating plant response to biotic stresses such as pathogens and insect attack.

5.3.2 *Melatonin and ROS*

In recent years, increasing evidence indicates that melatonin and ROS signal interact with each other to regulate multiple biological processes such as seed germination, stomatal movements, delay of leaf senescence, and stress defense. Li et al. (2016a) have shown that exogenous melatonin can induce cucumber defense against methyl viologen-induced oxidative stress through enhancing reduced/oxidized glutathione (GSH/GSSG) ratio in an H_2O_2-dependent manner, initiating research on the interaction between melatonin and ROS signal in regulating plant adaption to adverse environmental conditions. Gong et al. (2017) studied the crosstalk between melatonin and H_2O_2 in the regulation of stress tolerance in *Solanum lycopersicum*. They found that melatonin enhanced H_2O_2 accumulation by activating NADPH oxidase and induced stress tolerance. Inhibition of ROS generation or removal of H_2O_2 obviously alleviated the melatonin-mediated increase in the activities of antioxidant enzymes and the expression of some stress-related genes under stress conditions, revealing a novel mechanism in which RBOH-dependent H_2O_2 is an important component of the melatonin-induced plant tolerance against stress. Chen et al. (2017) provided genetic and molecular evidence that melatonin-induced reestablishment of ion and redox homeostasis and subsequent salinity tolerance is dependent *AtrbohF*-dependent ROS signaling. Consistently, Zhang et al. (2020) observed that melatonin-induced salt tolerance is related to H_2O_2 signaling in cucumber seedlings. When plants suffer cold stress, melatonin promotes H_2O_2 accumulation by upregulating the expression of *RBOHD* and $[Ca^{2+}]_{cyt}$ accumulation. H_2O_2 and Ca^{2+} promote each other to form a positive regulatory loop that is required for melatonin-mediated upregulation of CBF pathway-related genes and consequent induction of cold tolerance (Chang et al. 2021). In grafted watermelon plants, melatonin and MeJA form a reciprocal positive feedback loop to induce H_2O_2 generation and consequent cold tolerance (Li et al. 2021a). In addition to abiotic stress, melatonin also interacts with ROS to regulate plant response to biotic stress. Lee and Back (2017) noted that melatonin functions downstream of H_2O_2 in inducing plant resistance against *Pst* DC3000 infection.

Melatonin and ROS interaction in the regulation of plant growth and development also have been reported. Melatonin could not only enhance the level of endogenous H_2O_2, but also cause the induction of lateral root formation in alfalfa seedlings. Melatonin-modulated lateral root formation and related marker gene expression were obviously blocked or impaired after the scavenge of H_2O_2 in both transgenic *Arabidopsis* and alfalfa, suggesting that H_2O_2 functions as downstream of melatonin in the regulation of lateral root formation (Chen et al. 2018). Furthermore, melatonin facilitates lateral root formation by coordinating H_2O_2 and superoxide radicals that are derived by *polyamine oxidase* and *Rboh*, respectively (Chen et al. 2019). Li et al. (2021b) found that melatonin promoted seed germination under ABA stress is correlated with the increased accumulation of H_2O_2, which subsequently, promotes ABA catabolism and GA_3 biosynthesis and thus reduces the ratio of ABA/GA_3 balance. They also found that *RBOHD*-dependent H_2O_2 production is required for melatonin-delayed ABA-caused leaf senescence via triggering $[Ca^{2+}]_{cyt}$ accumulation, which can reduce K^+ efflux and cell death in response to ABA (Guo et al. 2023). Melatonin also regulates fruit metabolism via interaction with the H_2O_2. For example, melatonin could increase the level of H_2O_2 and promote berry ripening, whereas the application of an inhibitor of H_2O_2 generation inhibited melatonin-induced promotion of berry ripening, indicating that H_2O_2 participates in melatonin regulating berry ripening (Xu et al. 2018).

5.3.3 Melatonin and NO

Melatonin and NO interact with each other to regulate plant growth and development. In tomato seedlings, the application of exogenous melatonin at suitable concentrations, such as 50 μM, elevated the level of endogenous NO by inducing and reducing the expression of *NR* and *GSNOR*, respectively, and induced adventitious root formation. However, the effect of melatonin on adventitious root formation was abolished by *GSNOR*-overexpression or NO scavenger that reduce NO levels, indicating that NO participates in the melatonin-induced adventitious root formation and functions as a downstream signal (Wen et al. 2016). Mukherjee (2019) has reviewed the crosstalk between melatonin and NO during ethylene-mediated fruit ripening and drawn conclusions that melatonin mediates the NO and ethylene interaction via *N*-Nitrosomelatonin signaling in the regulation of fruit ripening. In postharvest tomato fruits, exogenous melatonin increased the activities of NO biosynthesis-related enzymes and endogenous NO accumulation, and upregulated the relative expression level of *NOS1* related to NO biosynthesis to confer chilling tolerance (Aghdam et al. 2019). In European pear (*Pyrus communis* L.), melatonin and NO reduce ethylene production, which mainly contributed to the ripening and softening processes, to delay postharvest senescence (Liu et al. 2019).

Melatonin increased the activity of NO biosynthesis by upregulating *PcNOS* expression and subsequently leads to the elevation of NO content. However, inhibition of NO biosynthesis virtually prevented melatonin-mediated delay of fruit senescence, demonstrating that melatonin acts on the upstream process of NO biosynthesis, thereby delaying the aging of pear fruits.

It has been evidenced that melatonin and NO function cooperatively to regulate plant response to various abiotic stresses. In tomato fruits exposed to chilling at 4 °C, exogenous application of melatonin-conferred tolerance is closely correlated with it-induced higher NO level which may arise from higher expression of *NOS* and higher activity of NOS enzyme (Aghdam et al. 2019). In sunflower seedlings with salt (120 mM NaCl) treatment, application of both melatonin and sodium nitroprusside (SNP, a NO donor) increased the ratio of GSH/GSSG by inducing GR activity and thus reducing the accumulation of ROS (Kaur and Bhatla 2016). SNP increased the activity of HIOMT, an important enzyme involved in melatonin biosynthesis, and subsequently promoted melatonin accumulation, while exogenous melatonin is unable to modulate the level of endogenous NO, highlighting a critical role of melatonin in NO-conferred salt tolerance. Similarly, Arora and Bhatla (2017) also found that NO functions as a positive regulator of melatonin accumulation in regulating sunflower seedling growth under salt stress. In pepper, NO participates in melatonin-enhanced salt tolerance via interacting with H_2S (Kaya et al. 2020). In rapeseed seedlings, exogenous melatonin elevated the level of NO and enhances NaCl tolerance by reestablishing redox and ion homeostasis; however, the positive effects of melatonin on NaCl tolerance were abolished by the removal of NO by NO scavenger and mutation of *NOA1* or *NIA1/2*, which deficiency exhibiting lower level of endogenous NO and null NR activity, respectively (Zhao et al. 2018). Moreover, scavenge of NO almost unaffected the level of endogenous melatonin in roots that are exposed to NaCl stress. These results indicate that NO acts as downstream of melatonin in enhancing NaCl tolerance. The study by Liu et al. (2015) revealed that melatonin application reduces Na^+ level, improves the antioxidant ability, and consequently increases alkaline tolerance by elevating NO level in tomato. In response to heat stress, the application of exogenous melatonin increases endogenous NO content accompanied by increases in the activities of NR and NOS and the expression of NR- and NOS-related genes, suggesting that NO may be involved in melatonin-induced heat tolerance of tomato seedlings (Jahan et al. 2019). Kaya et al. (2019, 2020) have studied the role of NO in melatonin-induced wheat seedling tolerance against cadmium (Cd) stress. Melatonin-triggered endogenous NO production and reduced oxidative stress, thus, enhancing Cd tolerance in wheat, while such enhancement by melatonin was prevented by the supply of NO scavengers, suggesting that NO plays a critical role in melatonin enhancing wheat tolerance against Cd toxicity. Consistently, Nabaei and Amooaghaie (2019) found that NO acts as a downstream signal participating in melatonin-induced antioxidant ability and subsequent Cd tolerance in *Catharanthus roseus*. Application of cPTIO, a NO scavenger, impaired Cd-induced melatonin biosynthesis, suggesting that NO is involved in the regulation of melatonin biosynthesis in rice response to Cd (Lee et al. 2017). Similarly, melatonin-induced maize seedling tolerance to lead (Pb);

however, such induction by melatonin was abolished by removal of NO by cPTIO, indicating that melatonin improves tolerance against Pb toxicity through NO signal in maize plants (Okant and Kaya 2019).

Melatonin application increases NO level and improves iron (Fe) deficiency tolerance in *Arabidopsis* (Zhou et al. 2016). Fe deficiency rapidly induces the biosynthesis of melatonin, which increases the soluble Fe contents in both shoots and roots, thus alleviating the chlorosis caused by Fe deficiency. Melatonin also significantly increases the NO accumulation in Fe-deficient roots, whereas, melatonin-induced Fe remobilization and reduction of leaf chlorosis is attenuated or blocked in the NO-deficient plants, indicating that melatonin increases plant tolerance to Fe-deficiency in a NO-dependent manner. Furthermore, Kaya et al. (2020) found that NO plays a critical role in melatonin-induced tolerance to iron deficiency through interaction with H_2S. In response to nitrogen starvation in *Haematococcus pluvialis*, melatonin confers tolerance through activating NO-mediated MAPK cascades and cyclic adenosine monophosphate (cAMP) signaling (Ding et al. 2018). Melatonin also interacts with NO to regulate plant immunity against the pathogen. Shi et al. (2015) found that infection of *Pst* DC3000 obviously promotes the accumulation of melatonin and NO, both of which enhances resistance to *Pst* DC3000 in *Arabidopsis* leaves. Application of exogenous melatonin enhanced the level of endogenous NO, whereas, the induction of innate immunity against *Pst* DC3000 infection by melatonin is inhibited or blocked by supply of NO scavenger or in *Arabidopsis* mutants with NO-deficiency. Therefore, the above results provide evidence that NO plays a critical role in melatonin inducing resistance against *Pst* DC3000 and acts as downstream. In *Arabidopsis* and tobacco, melatonin induces defense responses to pathogens through activation of MAPK kinase 4/5/7/9/ (MAPKK4/5/7/9)-mediated MAPK3/6 cascades (Lee and Back 2016). Furthermore, Lee and Back (2017) found that melatonin-triggered MAPKK kinase 3 (MAPKKK3) and downstream MAPK3/6 cascades is essential for NO-innated immunity response to *Pst* DC3000. In response to rice stripe virus (RSV), Lu et al. (2019) have demonstrated that melatonin can improve rice resistance to RSV through increasing the accumulation of NO. Also, NO mediates melatonin-induced plant resistance to tobacco mosaic virus in an SA-dependent pathway (Zhao et al. 2019).

5.3.4 Melatonin and H_2S

Several recent studies have reported the crosstalk between melatonin and H_2S in regulating plant response to abiotic stresses and H_2S may function downstream of melatonin. Heat reduces photosynthesis owing to the excessive ROS-caused oxidative stress and carbohydrate metabolism disturbances, to a certain extent. In wheat exposed to heat stress, exogenous melatonin increases the photosynthesis by reducing ROS and increasing carbohydrate metabolism; however, the application of H_2S scavenger (hypotaurine) abolishes the positive effect of melatonin, suggesting that

H_2S is required for melatonin-induced photosynthesis via enhancing carbohydrate metabolism in response to heat (Iqbal et al. 2021). In response to drought stress in *Arabidopsis*, exogenous melatonin upregulates the relative expression of H_2S biosynthesis-related genes *LCD* and *DES1* and consequently increased the accumulation of H_2S; however, melatonin-induced stomatal closure via activating K^+ channel and subsequent enhancement of drought tolerance is attenuated by knockout of *lcd*, *des1*, and *lcd/des1*, implying that melatonin induces stomatal closure and drought tolerance in a H_2S-dependent pathway (Wang et al. 2023). H_2S also is involved in melatonin-induced tolerances to salt stress as well as iron deficiency through interacting with NO and MAPK cascades and functions as downstream of melatonin (Kaya et al. 2020; Sun et al. 2021). Recently, Kaya et al. (2022) studied the crosstalk between melatonin and H_2S in regulating plant tolerance to arsenic toxicity. Melatonin induces arsenic tolerance and H_2S accumulation in pepper plants, while inhibition of H_2S generation by hypotaurine counteracts the beneficial effects of melatonin, demonstrating that H_2S is required for the tolerance to arsenic stress induced by melatonin. Under chromium toxicity, exogenous melatonin could maintain K^+ homeostasis and normal nitrogen metabolism, modulate ascorbate-glutathione (AsA-GSH) cycle, and reduce ROS level, electrolyte leakage, and lipid peroxidation. However, removal of H_2S by hypotaurine blocks the effect of melatonin on the above-mentioned parameters under chromium stress, suggesting that melatonin-induced alleviation of chromium-caused impairments in plants essentially involves endogenous H_2S (Khan et al. 2023). Although the crosstalk between melatonin and H_2S in regulating plant responses to some abiotic stresses has been reported; however, their relationships and mechanisms in regulating many physiological and biological processes need to be further studied.

5.4 Conclusions and Future Perspectives

Clearly, the action of melatonin becomes more evident and concise in the regulation of multiple physiological processes, when plants grow under normal conditions or stressful environments. Undoubtedly, plant hormones and many second messengers such as Ca^{2+}, ROS, NO, and H_2S, play important roles in the regulation of plant growth, development, and response to various environmental stresses. As has been seen throughout this chapter, melatonin interacts with plant hormones and many second messengers to regulate multiple physiological processes in plants (Fig. 5.1). Plant hormones and second messengers also interact with each other, forming complex regulatory networks.

However, it must be taken into account that only a small amount of molecular studies between melatonin and other signaling molecules have been carried out. Many more experiments are necessary, for example, using transgenic plants. There are many transgenic plant lines with deficiency or over accumulation of melatonin, plant hormones, and second messengers. Unfortunately, there are few studies with them. In addition, the interaction between melatonin and other signals in regulating

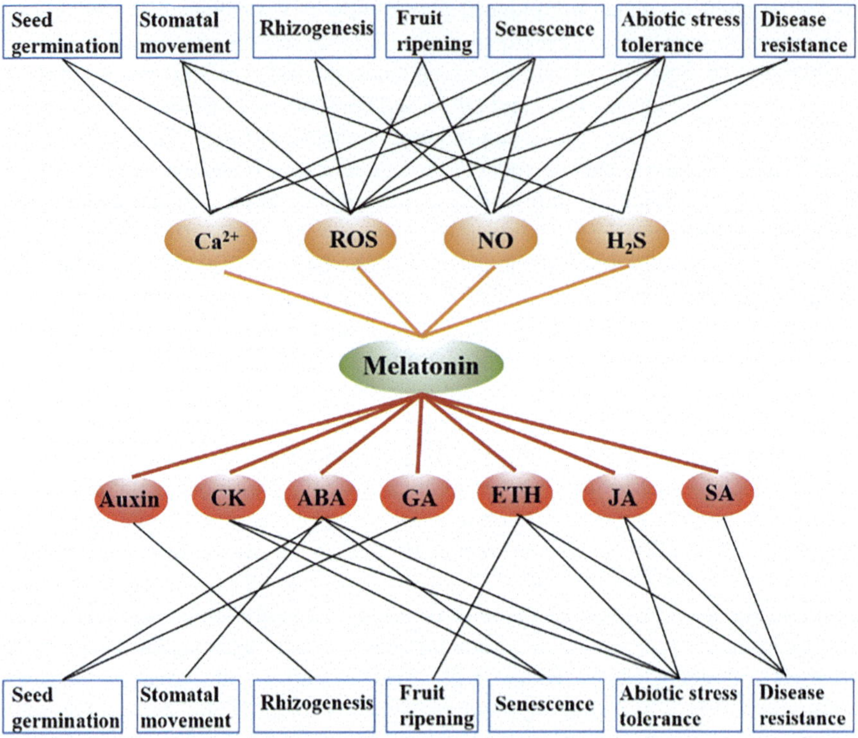

Fig. 5.1 Interaction between melatonin and other signaling molecules in plants

many physiological processes in plants, such as plant response to pest attack, and the underlying signal transduction and molecular mechanisms need to be further explored.

References

Aghdam MS, Luo Z, Jannatizadeh A, Sheikh-Assadi M, Sharafi Y, Farmani B, Fard JR, Razavi F (2019) Employing exogenous melatonin applying confers chilling tolerance in tomato fruits by upregulating ZAT2/6/12 giving rise to promoting endogenous polyamines, proline, and nitric oxide accumulation by triggering arginine pathway activity. Food Chem 275:549–556

Altaf MA, Shahid R, Ren MX, Mora-Poblete F, Arnao MB, Naz S, Anwar M, Altaf MM, Shahid S, Shakoor A, Sohail H, Ahmar S, Kamran M, Chen JT (2021) Phytomelatonin: an overview of the importance and mediating functions of melatonin against environmental stresses. Physiol Plant 172:820–846

Arnao MB, Hernández Ruiz J (2015) Functions of melatonin in plants: a review. J Pineal Res 59: 133–150

Arnao MB, Hernández-Ruiz J (2007) Inhibition of ACC oxidase activity by melatonin and IAA in etiolated lupin hypocotyls. In: Ramina A, Chang C, Giovannoni J, Klee H, Perata P, Woltering E (eds) Advances in plant ethylene research. Springer, Dordrecht, pp 101–103

Arnao MB, Hernández-Ruiz J (2009) Protective effect of melatonin against chlorophyll degradation during the senescence of barley leaves. J Pineal Res 46:58–63

Arnao MB, Hernandez-Ruiz J (2014) Melatonin: plant growth regulator and/or biostimulator during stress? Trends Plant Sci 19:789–797

Arnao MB, Hernández-Ruiz J (2017) Growth activity, rooting capacity, and tropism: three auxinic precepts fulfilled by melatonin. Acta Physiol Plant 39:127

Arnao MB, Hernández-Ruiz J (2018a) Melatonin: a new plant hormone and/or a plant master regulator? Trends Plant Sci 24:38–48

Arnao MB, Hernández-Ruiz J (2018b) Melatonin and its relationship to plant hormones. Ann Bot 121:195–207

Arora D, Bhatla SC (2017) Melatonin and nitric oxide regulate sunflower seedling growth accompanying differential expression of Cu/Zn SOD and Mn SOD. Free Radic Biol Med 106:315–328

Baxter A, Mittler R, Suzuki N (2014) ROS as key players in plant stress signalling. J Exp Bot 65: 1229–1240

Berridge MJ, Lipp P, Bootman MD (2000) The versatility and universality of calcium signalling. Nat Rev Mol Cell Biol 1:11–21

Besson-Bard A, Pugin A, Wendehenne D (2008) New insights into nitric oxide signaling in plants. Annu Rev Plant Biol 59:21–39

Bose J, Rodrigo-Moreno A, Shabala S (2014) ROS homeostasis in halophytes in the context of salinity stress tolerance. J Exp Bot 65:1241–1257

Chang JJ, Guo YL, Li JY, Su ZZ, Wang CX, Zhang RM, Wei CH, Ma JX, Zhang X, Li H (2021) Positive interaction between H_2O_2 and Ca^{2+} mediates melatonin-induced CBF pathway and cold tolerance. Antioxidants 10:1451

Chen ZP, Xie YJ, Gu Q, Zhao G, Zhang YH, Cui WT, Xu S, Wang R, Shen WB (2017) The *AtrbohF*-dependent regulation of ROS signaling is required for melatonin-induced salinity tolerance in *Arabidopsis*. Free Radic Biol Med 108:465–477

Chen ZP, Gu Q, Yu XL, Huang LQ, Xu S, Wang R, Shen W, Shen WB (2018) Hydrogen peroxide acts downstream of melatonin to induce lateral root formation. Ann Bot 121:1127–1136

Chen J, Li H, Yang K, Wang YZ, Yang LF, Hu LB, Liu RX, Shi ZQ (2019) Melatonin facilitates lateral root development by coordinating PAO-derived hydrogen peroxide and Rboh-derived superoxide radical. Free Radic Biol Med 143:534–544

Dat J, Vandenabeele S, Vranova E, Montagu MV, Inzé D, Breusegem FV (2000) Dual action of the active oxygen species during plant stress responses. Cell Mol Life Sci 57:779–795

Ding W, Zhao Y, Xu JW (2018) Melatonin: a multifunctional molecule that triggers defense responses against high light and nitrogen starvation stress in *Haematococcus pluvialis*. J Agric Food Chem 66:7701–7711

Ding F, Ren LM, Xie F, Wang ML, Zhang SX (2022) Jasmonate and melatonin act synergistically to potentiate cold tolerance in tomato plants. Front Plant Sci 12:763284

Dodd AN, Kudla J, Sanders D (2010) The language of calcium signaling. Annu Rev Plant Biol 61: 593–620

Fan JB, Xie Y, Zhang ZC, Chen L (2018) Melatonin: a multifunctional factor in plants. Int J Mol Sci 19:1528

Fang PP, Wang Y, Wang MQ, Wang F, Chi C, Zhou YH, Zhou J, Shi K, Xia XJ, Foyer CH, Yu JQ (2021) Crosstalk between brassinosteroid and redox signaling contributes to the activation of CBF expression during cold responses in tomato. Antioxidants 10:509

Footitt S, Douterelo-Soler I, Clay H, Finch-Savage WE (2011) Dormancy cycling in *Arabidopsis* seeds is controlled by seasonally distinct hormone-signaling pathways. Proc Natl Acad Sci U S A 108:20236–20241

Fotopoulos V, Christou A, Antoniou C, Manganaris GA (2015) Hydrogen sulphide: a versatile tool for the regulation of growth and defence responses in horticultural crops. J Hortic Sci Biotechnol 90:227–234

Fu J, Wu Y, Miao Y et al (2017) Improved cold tolerance in *Elymus nutans* by exogenous application of melatonin may involve ABA-dependent and ABA-independent pathways. Sci Rep 7:39865

Gong B, Yan YY, Wen D, Shi QH (2017) Hydrogen peroxide produced by NADPH oxidase: a novel downstream signaling pathway in melatonin-induced stress tolerance in *Solanum lycopersicum*. Physiol Plant 160:396–409

Guo HM, Xiao TY, Zhou H, Xie YJ, Shen WB (2016) Hydrogen sulfide: a versatile regulator of environmental stress in plants. Acta Physiol Plant 38:16

Guo YL, Yan JY, Su ZZ, Chang JJ, Yang JX, Wei CH, Zhang Y, Ma JX, Zhang, Li H (2021) Abscisic acid mediates grafting-induced cold tolerance of watermelon via interaction with melatonin and methyl jasmonate. Front Plant Sci 12:785317

Guo YL, Zhu JY, Liu JH, Xue YX, Chang JJ, Zhang Y, Ahammed GJ, Wei CH, Ma JX, Li PF, Zhang X, Li H (2023) Melatonin delays ABA-induced leaf senescence via H_2O_2-dependent calcium signalling. Plant Cell Environ 46:171–184

Imran M, Khan AL, Mun BG, Bilal S, Shaffique S, Kwon EH, Kang SM, Yun BW, Lee IJ (2022) Melatonin and nitric oxide: dual players inhibiting hazardous metal toxicity in soybean plants via molecular and antioxidant signaling cascades. Chemosphere 308:136575

Iqbal N, Fatma M, Gautam H, Umar S, Sofo A, D'ippolito I, Khan NA (2021) The crosstalk of melatonin and hydrogen sulfide determines photosynthetic performance by regulation of carbohydrate metabolism in wheat under heat stress. Plan Theory 10:1778

Jahan MS, Shu S, Wang Y, Chen Z, He MM, Tao MQ, Sun J, Guo SR (2019) Melatonin alleviates heat-induced damage of tomato seedlings by balancing redox homeostasis and modulating polyamine and nitric oxide biosynthesis. BMC Plant Biol 19:414

Kaur H, Bhatla SC (2016) Melatonin and nitric oxide modulate glutathione content and glutathione reductase activity in sunflower seedling cotyledons accompanying salt stress. Nitric Oxide 59: 42–53

Kaya C, Okant M, Ugurlar F, Alyemeni MN, Ashraf M, Ahmad P (2019) Melatonin-mediated nitric oxide improves tolerance to cadmium toxicity by reducing oxidative stress in wheat plants. Chemosphere 225:627–638

Kaya C, Higgs D, Ashraf M, Alyemenid MN, Ahmad P (2020) Integrative roles of nitric oxide and hydrogen sulfide in melatonin-induced tolerance of pepper (*Capsicum annuum* L.) plants to iron deficiency and salt stress alone or in combination. Physiol Plant 168:256–277

Kaya C, Ugurlar F, Ashraf M, Alyemeni MN, Bajguz A, Ahmad P (2022) The involvement of hydrogen sulphide in melatonin-induced tolerance to arsenic toxicity in pepper (*Capsicum annuum* L.) plants by regulating sequestration and subcellular distribution of arsenic, and antioxidant defense system. Chemosphere 309:136678

Khan MN, Siddiqui MH, Mukherjee S, AlSolami MA, Alhussaen KM, AlZuaibr FM, Siddiqui ZH, Al-Amri AA, Alsubaie QD (2023) Melatonin involves hydrogen sulfide in the regulation of H^+-ATPase activity, nitrogen metabolism, and ascorbate-glutathione system under chromium toxicity. Environ Pollut 323:121173

Kim JE, Kim WT, Kang BG (2001) IAA and N6-benzyladenine inhibit ethylene-regulated expression of ACC oxidase and ACC synthase genes in mungbean hypocotyls. Plant Cell Physiol 42: 1056–1061

Kunkel BN, Brooks DM (2002) Cross talk between signaling pathways in pathogen defense. Curr Opin Plant Biol 5:325–331

Lee HY, Back K (2016) Mitogen-activated protein kinase pathways are required for melatonin-mediated defense responses in plants. J Pineal Res 60:327–335

Lee HY, Back K (2017) Melatonin is required for H_2O_2—and NO-mediated defense signaling through MAPKKK3 and OXI1 in *Arabidopsis thaliana*. J Pineal Res 62:e12379

Lee HY, Byeon Y, Back K (2014) Melatonin as a signal molecule triggering defense responses against pathogen attack in *Arabidopsis* and tobacco. J Pineal Res 57:262–268

Lee HY, Byeon Y, Tan DX, Reiter RJ, Back K (2015) *Arabidopsis* serotonin N-acetyltransferase knockout mutant plants exhibit decreased melatonin and salicylic acid levels resulting in susceptibility to an avirulent pathogen. J Pineal Res 58:291–299

Lee K, Choi GH, Back K (2017) Cadmium-induced melatonin synthesis in rice requires light, hydrogen peroxide, and nitric oxide: key regulatory roles for tryptophan decarboxylase and caffeic acid *O*-methyltransferase. J Pineal Res 63:e12441

Li ZG (2015) Analysis of some enzyme activities of hydrogen sulfide metabolism in plants. Methods Enzymol 555:253–269

Li C, Liang D, Chang C, Jia D, Ma F (2015) Melatonin mediates the regulation of ABA metabolism, free-radical scavenging, and stomatal behavior in two Malus species under drought stress. J Exp Bot 66:669–680

Li H, He J, Yang XZ, Li X, Luo D, Wei CH, Ma JX, Zhang Y, Yang JQ, Zhang X (2016a) Glutathione-dependent induction of local and systemic defense against oxidative stress by exogenous melatonin in cucumber (*Cucumis sativus* L.). J Pineal Res 60:206–216

Li XN, Tan DX, Jiang D, Liu FL (2016b) Melatonin enhances cold tolerance in drought-primed wild-type and abscisic acid-deficient mutant barley. J Pineal Res 61:328–339

Li H, Chang JJ, Zheng JX, Dong YC, Liu QY, Yang XZ, Wei CH, Zhang Y, Ma JX, Zhang X (2017) Local melatonin application induces cold tolerance in distant organs of *Citrullus lanatus* L. via long distance transport. Scientific. Report 7:e40858

Li H, Guo YL, Lan ZX, Xu K, Chang JJ, Ahammed GJ, Ma JX, Wei CH, Zhang X (2021a) Methyl jasmonate mediates melatonin-induced cold tolerance of grafted watermelon plants. Hortic Res 8:57

Li H, Guo YL, Lan ZX, Zhang ZX, Ahammed GJ, Chang JJ, Zhang Y, Wei CH, Zhang X (2021b) Melatonin antagonizes ABA action to promote seed germination by regulating Ca^{2+} efflux and H_2O_2 accumulation. Plant Sci 303:110761

Li SG, Huan C, Liu Y, Zheng XL, Bi Y (2022) Melatonin induces improved protection against *Botrytis cinerea* in cherry tomato fruit by activating salicylic acid signaling pathway. Sci Hortic 304:111299

Liang C, Li A, Yu H, Li WA, Liang CZ, Guo SD, Zhang R, Chu CC (2017) Melatonin regulates root architecture by modulating auxin response in rice. Front Plant Sci 8:134

Liu N, Gong B, Jin Z, Wang X, Wei M, Yang F, Li Y, Shi Q (2015) Sodic alkaline stress mitigation by exogenous melatonin in tomato needs nitric oxide as a downstream signal. J Plant Physiol 186–187:68–77

Liu JL, Yang J, Zhang HQ, Cong L, Zhai R, Yang CQ, Wang ZG, Ma FW, Xu LF (2019) Melatonin inhibits ethylene synthesis via nitric oxide regulation to delay postharvest senescence in pears. J Agric Food Chem 67:2279–2288

Liu QP, Atta UR, Wang FR, Liu KX, Ma X, Weng QB (2021) Defense-related hormone signaling coordinately controls the role of melatonin during *Arabidopsis thaliana-Pseudomonas syringae* interaction. Eur J Plant Pathol 160:707–716

Lozano-Juste J, León J (2010) Enhanced abscisic acid mediated responses in *nia1nia2noa1-2* triple mutant impaired in NIA/NR- and AtNOA1-dependent nitric oxide biosynthesis in *Arabidopsis*. Plant Physiol 152:891–903

Lu RF, Liu ZY, Shao YD, Sun F, Zhang YL, Cui J, Zhou YJ, Shen WB, Zhou T (2019) Melatonin is responsible for rice resistance to rice stripe virus infection through a nitric oxide-dependent pathway. Virol J 16:141

Ma XQ, Zhang J, Burgess P, Rossi S, Huang B (2018) Interactive effects of melatonin and cytokinin on alleviating drought-induced leaf senescence in creeping bentgrass (*Agrostis stolonifera*). Environ Exp Bot 145:1–11

Ma C, Pei ZQ, Bai X, Lu SH, Su M, Kang X, Gao HX, Wang J, Zhang TG, Zheng S (2022) Exogenous melatonin and $CaCl_2$ alleviate cold-induced oxidative stress and photosynthetic inhibition in cucumber seedlings. J Plant Growth Regul 42:1–18. https://doi.org/10.1007/s00344-022-10805-z

Mukherjee S (2019) Recent advancements in the mechanism of nitric oxide signaling associated with hydrogen sulfide and melatonin crosstalk during ethylene-induced fruit ripening in plants. Nitric Oxide 82:25–34

Mur LAJ, Mandon J, Persijn S, Cristescu SM, Moshkov IE, Novikova GV, Hall MA, Harren FJM, Hebelstrup KH, Gupta KJ (2012) Nitric oxide in plants: an assessment of the current state of knowledge. AoB Plants 5:pls052

Nabaei M, Amooaghaie R (2019) Nitric oxide is involved in the regulation of melatonin-induced antioxidant responses in *Catharanthus roseus* roots under cadmium stress. Botany 97:681–690

Okant M, Kaya C (2019) The role of endogenous nitric oxide in melatonin-improved tolerance to lead toxicity in maize plants. Environ Sci Pollut Res 26:11864–11874

Pandey AK, Gautam A (2020) Stress responsive gene regulation in relation to hydrogen sulfide in plants under abiotic stress. Physiol Plant 2:511–525

Shan SS, Wang ZQ, Pu HL, Duan WH, Song HM, Li JK, Zhang ZK, Xu XB (2022) DNA methylation mediated by melatonin was involved in ethylene signal transmission and ripening of tomato fruit. Sci Hortic 291:110566

Sharma A, Zheng B (2019) Melatonin mediated regulation of drought stress: physiological and molecular aspects. Plan Theory 8(7):190

Sharma A, Wang J, Xu D, Tao S, Chong S, Yan D et al (2020) Melatonin regulates the functional components of photosynthesis, antioxidant system, gene expression, and metabolic pathways to induce drought resistance in grafted Carya cathayensis plants. Sci Total Environ 713:136675

Shi HT, Chen YH, Tan DX, Reiter RJ, Chan ZL, He CZ (2015) Melatonin induces nitric oxide and the potential mechanisms relate to innate immunity against bacterial pathogen infection in *Arabidopsis*. J Pineal Res 59:102–108

Skinner JS, von Zitzewitz J, Szucs P, Marquez-Cedillo L, Filichkin T, Amundsen K, Stockinger EJ, Thomashow MF, Chen THH, Hayes PM (2005) Structural, functional, and phylogenetic characterization of a large *CBF* gene family in barley. Plant Mol Biol 59:533–551

Sun QQ, Zhang N, Wang JF, Zhang HJ, Li DB, Shi J, Li R, Weeda S, Zhao B, Ren SX, Guo YD (2015) Melatonin promotes ripening and improves quality of tomato fruit during postharvest life. J Exp Bot 66:657–668

Sun QQ, Zhang N, Wang JF, Cao YY, Li XS, Zhang HJ, Zhang L, Tan DX, Guo YD (2016) A label-free differential proteomics analysis reveals the effect of melatonin on promoting fruit ripening and anthocyanin accumulation upon postharvest in tomato. J Pineal Res 61:138–153

Sun YP, Ma C, Kang X, Zhang L, Wang J, Zheng S, Zhang TG (2021) Hydrogen sulfide and nitric oxide are involved in melatonin-induced salt tolerance in cucumber. Plant Physiol Biochem 167:101–112

Tian X, He XD, Xu JP, Yang ZF, Fang WM, Yin YQ (2022) Mechanism of calcium in melatonin enhancement of functional substance-phenolic acid in germinated hulless barley. RSC Adv 12: 29214

Vafadar F, Amooaghaie R, Ehsanzadeh P, Ghanadian M, Talebi M, Ghanati F (2020a) Melatonin and calcium modulate the production of rosmarinic acid, luteolin, and apigenin in *Dracocephalum kotschyi* under salinity stress. Phytochemistry 177:112422

Vafadar F, Amooaghaie R, Ehsanzadeh P, Ghanati F, Sajedi RH (2020b) Crosstalk between melatonin and Ca^{2+}/CaM evokes systemic salt tolerance in *Dracocephalum kotschyi*. J Plant Physiol 252:153237

Vafadar F, Amooaghaie R, Ehsanzadeh P, Ghanati F, Allakhverdiev S (2021) Melatonin improves the photosynthesis in *Dracocephalum kotschyi* under salinity stress in a Ca^{2+}/CaM-dependent manner. Funct Plant Biol 49:89–101

Wang Q, An B, Wei Y, Reiter RJ, Shi HT, Luo HL, He CZ (2016) Melatonin regulates root meristem by repressing auxin synthesis and polar auxin transport in *Arabidopsis*. Front Plant Sci 7:1882

Wang ZQ, Mu Y, Zhang LP, Liu ZQ, Liu DM, Jin ZP, Pei YX (2023) Hydrogen sulfide mediated the melatonin induced stoma closure by regulating the K^+ channel in *Arabidopsis thaliana*. Environ Exp Bot 205:105125

Weeda S, Zhang N, Zhao XL, Ndip G, Guo YD, Buck GA, Fu CG, Ren SX (2014) *Arabidopsis* transcriptome analysis reveals key roles of melatonin in plant defense systems. PLoS One 9: e93462

Wei J, Li DX, Zhang JR, Shan C, Rengel Z, Song ZB, Chen Q (2018) Phytomelatonin receptor PMTR1-mediated signaling regulates stomatal closure in *Arabidopsis thaliana*. J Pineal Res 65: e12500

Wen D, Gong B, Sun SS, Liu SQ, Wang XF, Wei M, Yang FJ, Li Y, Shi QH (2016) Promoting roles of melatoninin adventitious root development of *Solanum lycopersicum* L. by regulating auxin and nitric oxide signaling. Front Plant Sci 7:718

Xu LL, Yue QY, Xiang GQ, Bian FE, Yao YX (2018) Melatonin promotes ripening of grape berry via increasing the levels of ABA, H_2O_2, and particularly ethylene. Hortic Res 5:41

Xu LL, Xiang GQ, Sun QH, Ni Y, Jin ZX, Gao SW, Yao YX (2019) Melatonin enhances salt tolerance by promoting MYB108A-mediated ethylene biosynthesis in grapevines. Hortic Res 6: 114

Xuan LJ, Li J, Wang XY, Wang CY (2020) Crosstalk between hydrogen sulfide and other signal molecules regulates plant growth and development. Int J Mol Sci 21:4593

Yin LH, Wang P, Li MJ, Ke XW, Li CY, Liang D, Wu S, Ma XL, Li C, Zou YJ, Ma FW (2013) Exogenous melatonin improves Malus resistance to Marssonina apple blotch. J Pineal Res 54: 426–434

Zhang HM, Zhang YQ (2014) Melatonin: a well-documented antioxidant with conditional pro-oxidant actions. J Pineal Res 57:131–146

Zhang N, Sun QQ, Zhang HJ, Cao YY, Weeda S, Ren SX, Guo YD (2014a) Roles of melatonin in abiotic stress resistance in plants. J Exp Bot 66:647–656

Zhang HJ, Zhang N, Yang RC, Wang L, Sun QQ, Li DB, Cao YY, Weeda S, Zhao B, Ren SX, Guo YD (2014b) Melatonin promotes seed germination under high salinity by regulating antioxidant systems, ABA and GA_4 interaction in cucumber (*Cucumis sativus* L.). J Pineal Res 57:269–279

Zhang J, Shi Y, Zhang X, Du H, Xu B, Huang B (2017a) Melatonin suppression of heat-induced leaf senescence involves changes in abscisic acid and cytokinin biosynthesis and signaling pathways in perennial ryegrass (*Lolium perenne* L.). Environ Exp Bot 138:36–45

Zhang R, Sun Y, Liu Z, Jin W, Sun Y (2017b) Effects of melatonin on seedling growth, mineral nutrition, and nitrogen metabolism in cucumber under nitrate stress. J Pineal Res 62:e12403

Zhang TG, Shi ZF, Zhang XH, Zheng S, Wang J, Mo JG (2020) Alleviating effects of exogenous melatonin on salt stress in cucumber. Sci Hortic 262:109070

Zhao G, Zhao Y, Yu X, Kiprotich F, Han H, Guan R, Wang R, Shen W (2018) Nitric oxide is required for melatonin enhanced tolerance against salinity stress in rapeseed (*Brassica napus* L.) seedlings. Int J Mol Sci 19:1912

Zhao L, Chen L, Gu P, Zhan X, Zhang Y, Hou C, Wu Z, Wu YF, Wang QC (2019) Exogenous application of melatonin improves plant resistance to virus infection. Plant Pathol 68:1287–1295

Zhou C, Liu Z, Zhu L, Zhongyou Ma ZY, Wang JF, Zhu J (2016) Exogenous melatonin improves plant iron deficiency tolerance via increased accumulation of polyamine-mediated nitric oxide. Int J Mol Sci 17:1777

Zuo BX, Zheng XD, He PL, Wang L, Lei Q, Feng C, Zhou JZ, Li QT, Han ZH, Kong J (2014) Overexpression of *MzASMT* improves melatonin production and enhances drought tolerance in transgenic *Arabidopsis thaliana* plants. J Pineal Res 57:408–417

Chapter 6
Melatonin in Safeguarding Plants from High Light and UV Exposure

Ashish Sharma, Neha Verma, Harleen Kaur, Ravneet Kaur, Neetu Jagota, and Swapnil Singh

Abstract Although light is indispensable for plants but when it increases beyond a certain level it can result in damage to the photosynthetic machinery of the plants. An added threat to high light is the presence of UV radiations accompanying high light conditions. Plants have developed many mechanisms for the prevention of the photosynthetic apparatus and maintenance of growth and development under such conditions. An important determinant in this protection process is melatonin. A low molecular weight indoleamine, first identified from animals, is ubiquitously found in plants and is involved in phytoprotection under different adverse environmental conditions. Here we explore the involvement of this molecule in the protection of plants against light stress and establish a possible relationship between the variable luminous conditions and melatonin content in the plants.

Keywords Melatonin · High light · ROS · Photosynthesis · Variable environment

6.1 Introduction

Environmental stress can be considered as an inherent part of the plant's life cycle, as being sessile organisms they cannot exercise much choice in selecting or rejecting the different conditions available to them based on their comfort level. To survive and thrive in these limitless variable conditions that they have to experience every day, plants have developed multiple response mechanisms to the environmental conditions. In this respect after its first discovery in the plants in 1995 (Dubbels et al. 1995; Hattori et al. 1995) melatonin has been extensively researched for its involvement in variety of plant physiological processes including chloroplast biology,

A. Sharma (✉) · H. Kaur · R. Kaur · N. Jagota · S. Singh
Department of Botany and Environment Science, DAV University, Jalandhar, Punjab, India
e-mail: ashish10210@davuniversity.org

N. Verma
Government College of Education, Jalandhar, Punjab, India

A. Sharma, G. J. Ahammed (eds.), *Melatonin in Plants: Role in Plant Growth, Development, and Stress Response*, Plant Life and Environment Dynamics,
https://doi.org/10.1007/978-981-99-8051-2_6

maintenance of photosynthesis, plant morphogenesis to name some of the processes. It has also been shown to be involved in a variety of abiotic stress responses allowing the plants reproductive success even in the stressed conditions (Zeng et al. 2022). Melatonin has been shown to be present in a variety of plant parts, viz. leaves, roots, stems, petals, flower buds, fruits, and seeds of different plant species in which it has been studied. It has been shown to be synthesized in chloroplasts and mitochondria (Tan and Reiter 2020).

Sunlight is the absolute overall energy source for all organisms that include plants also and as such the evidence for the statement comes from the variety in which the protein complexes are found in plants that help to capture this light for photosynthesis. Plants mitigate low light conditions by rearranging the chloroplasts so as to capture more sunlight and maintain the necessary energy driven reactions. However, during many situations injury to the photosynthetic machinery has been observed when the available light surpasses the plants requirement which is the root cause of limiting productivity under light condition. Another major effect of high light conditions is the sudden increase in the reactive oxygen species (ROS) which time and again have been proven to be toxic to the plants.

Found throughout the plant kingdom, melatonin is an amphiphilic indoleamine that shows involvement in many physiological processes. Evidence generated by a number of workers over the years place melatonin in the category of growth promoting and root inducing molecule (Arnao and Hernández-Ruiz 2007; Hernández-Ruiz and Arnao 2008; Sarrou et al. 2014; Zhang et al. 2014). However, apart from developmental effects melatonin also exhibit defensive role in the plants against both biotic and abiotic stresses. It has been shown that plant species that produce more melatonin can better adapt to the prevailing stressful conditions (Zhang et al. 2015). Along with the naturally produced melatonin, exogenous application has also been found equally effective in protecting plants in the stressful environments. When present in sufficient concentration, melatonin can protect plants against a variety of abiotic stressors such as temperature extremes, radiations, salinity, drought, basically all those conditions that lead to the production of reactive oxygen species (ROS). Under stress conditions one of the major roles of melatonin is in the transitional developmental stages of flowering and fruiting. Therefore, apart from strengthening the defensive traits of the plants, the presence of melatonin is also important for developmental cues of the plants under environmental stress.

The cellular melatonin concentration greatly depends on the external environmental conditions and can get easily affected in response to a small change in available light, ambient temperature, and other stressors. It has also been reported that ambient environment of the plants can affect the endogenous melatonin concentration. As reported by Tan et al. (2007) the plants grown in the open field conditions produce almost 2.5 times more melatonin as compared to the plants grown in the artificial light conditions, clearly suggesting the role of prevalent environmental conditions in the accumulation of melatonin in the plants. Similar results were demonstrated in the pepper fruits where it was shown that melatonin content of the shaded fruits was lower as compared to fruits exposed to sunlight (Riga et al. 2014). Apart from the solar radiations, UV radiations can also influence the melatonin content of the exposed plants or plant parts. It was shown by Simopoulos et al. (2005) that a higher melatonin content accumulates in the plants exposed to more UV light as compared to the plants exposed to less UV light. Therefore, it may not be wrong to say that in plants presence of melatonin can have an important role in the process of photosynthesis.

6.2 Biosynthesis and Utilization

Melatonin is a universal molecule found in all animal and plant species (Wu et al. 2021). However, in both plants and animals the pathway leading to the synthesis of melatonin differ in the basic requirement of the enzymes. Where in animals, synthesis of melatonin follows a pathway involving four enzymes, in plants it is a six-enzyme pathway leading to the synthesis of melatonin molecule. However, in plants there exist two different pathways for the synthesis of melatonin that dominate in different conditions (Zeng et al. 2022). Under standard conditions, 5-hydroxy tryptamine produced in the chloroplasts as a result of the action of Tryptophan decarboxylase (TDC) is acted upon by serotonin N-acetyltransferase (SNAT) followed by either N-acetylserotonin-O-methyltransferase (ASMT) or caffeic acid O-methyltransferase (COMT) to give rise to melatonin with N-acetyl-5-hydroxy-tryptamine as an intermediate. Whereas under abiotic stress conditions, ASMT/COMT are first to act to produce 5-methoxy tryptamine followed by the action of SNAT to give rise to melatonin (Fig. 6.1). Although, the exact trigger for the switching of the pathways is yet to be explained but sufficient evidence has been generated over the years to account for the existence of the above mentioned two pathways (Ye et al. 2019; Tan and Reiter 2020).

As far as degradation of melatonin is concerned it can take either enzymatic route or non-enzymatic route which involves degradation of the molecule by the action of free radicals and/or oxidants (Sun et al. 2021). The enzymatic breakdown of melatonin has been shown to be different in different plant species and different parts of the plants. For example, enzyme indolamine 2,3-dioxygenase, identified from water hyacinth converts melatonin to N1-acetyl-N2-formyl-5-methoxykynuramine (Tan et al. 2007). Other enzymes that have been found to degrade melatonin include melatonin 2-hydroxylase and melatonin 3-hydroxylase commonly found in higher plants (Byeon and Back 2015; Lee et al. 2016). A dioxygenase isoform has been identified from rice chloroplasts which is an important enzyme in one of the most important metabolic pathways of melatonin under both biotic and abiotic stresses (Sun et al. 2021). As far as non-enzymatic conversion of melatonin is concerned some important reactions has been identified that can change the chemical nature of melatonin. One such reaction is nitrosation of the indole ring in melatonin giving rise to nitrosomelatonin (Zeng et al. 2022) rendering it non-functional. Several melatonin metabolites have been identified from maize and cucumber which form as a result of reaction of melatonin with ROS and reactive nitrogen species (RNS) (Kołodziejczyk et al. 2015; Manchester et al. 2015).

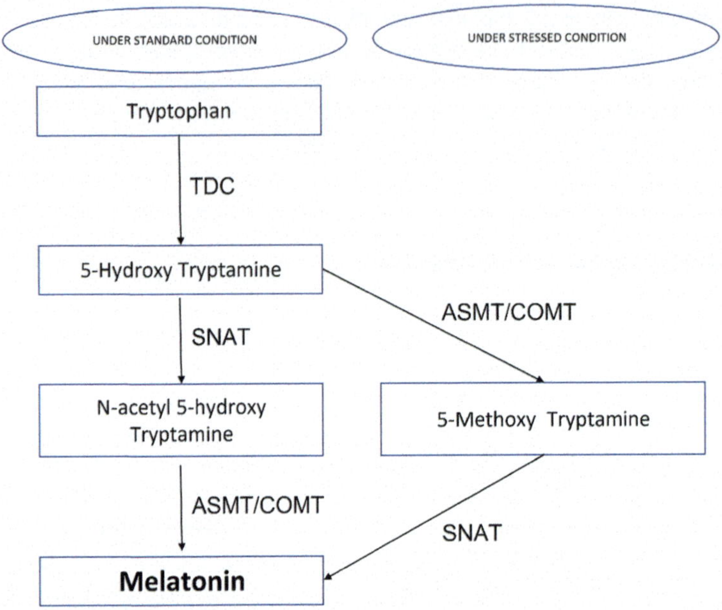

Fig. 6.1 Proposed synthesis of melatonin in chloroplasts under stressed and unstressed conditions. The abbreviations expanded as: *TDC* tryptophan decarboxylase, *SNAT* serotonin N-acetyltransferase, *ASMT* N-acetylserotonin-O-methyltransferase, *COMT* caffeic acid O-methyltransferase

6.3 Melatonin Under Light Stress

Light is an almost absolute necessity to the plants as it is required for one of the most important processes in the living world and also help in regulation of many morphogenetic processes in the plants. Apart from its incredible utility light can also be an environmental stressor when it reaches beyond a certain limit. At which time it becomes a primary source of photodamage and can also lead to overproduction of ROS in exposed plants. Another major problem that plants face is the presence of UV light that accompanies high light in the environment which leads to generation of excessive free radicals in plants (Nazir et al. 2020). Apart from generating large number of ROS, exposure of plants to UV light can also have harmful effects on the ultrastructure of plant cellular components and nuclear membrane.

Light can be considered as a major regulator of melatonin in different aspects of its synthesis and metabolism in plants. It can regulate the function of enzymes involved in the biosynthesis of the molecule such as tryptophan decarboxylase or it can also influence the enzymes involved in the breakdown of melatonin. Apart from its biosynthesis and degradation, evidence has also been generated for light being involved in the melatonin mediated defensive responses in the plants.

Given all the effects that light exhibit on plant it is also an important regulator of melatonin biosynthesis in plants (Zhang et al. 2019). Exposure of plants to UV radiations has been associated with induction of melatonin synthesis in plants. This increase in melatonin concentration in presence of UV radiations is evidentiary enough of the involvement of melatonin in UV light response in plants (Zeng et al. 2022). A marked difference in the melatonin levels has also been documented in the plants growing under high UV radiations when compared to the plants growing under low UV radiation in alpine and Mediterranean plant populations (Conti et al. 2002). In the experiments with water hyacinth, it was found that the plants growing in the open sunlight accumulate more melatonin in comparison to the plants that are propagated under artificial conditions. Analysis of the results suggested a possible relation in the growing conditions and melatonin content in the plants under study (Tan et al. 2007). In other experiments also it has been shown that plants that are grown under variable environmental conditions in the open field always accumulate more melatonin than the plants that grow in the controlled environment. This observation is consistent with the role that melatonin plays in the protection of plants and their photosynthetic machinery under variable light conditions. These findings have been confirmed in tomato plants grown indoors and outdoors (Arnao and Hernández-Ruiz 2013). It was reported that the plants that received moderate conditions in the indoor setting showed almost ten times lower melatonin accumulation in comparison to the plants that received full sunlight outdoors.

6.4 Melatonin and Photosynthesis

Photosynthesis is one of the most important physiological processes on this earth as it happens to be one and only source of entry of carbon into the living world, which is of utmost importance owing to the fact that life on earth is primarily carbon-based form. It is light that regulates this process primarily as it is the light energy that is used to move electrons in light reaction of photosynthesis to generate ATP, viz. light energy gets converted to chemical energy. But more often than not there is available excessive light in the environment. Also the prevalence of adverse environmental conditions such as drought, saline conditions, low or high temperature can create such conditions where normal light can also act as excessive light for the plants due to damage caused to the internal structure by different stressors. Therefore, protection of the photosynthetic machinery becomes extremely important for sustenance of plant functions. Here we present evidence for the involvement of melatonin in prevention of photosynthesis in plants.

One of the major effects of high light is the curling of leaves and degradation of chlorophyll molecule. Yang et al. 2021 showed that if melatonin is used as a pre-treatment before the onset of high light conditions, it can lead to an increase in the concentration of endogenous melatonin by almost 111% and also help the plants to overcome the damage and discoloration of the leaves. Exogenous melatonin can also improve the pigment component of the treated plants when used at the appropriate time. It was also shown that chlorophyll fluorescence parameters also responded to the presence of melatonin during high light conditions. Although chlorophyll fluorescence, photosynthetic rate and stomatal conductance showed a decline in presence of high light but in the plants treated with melatonin all these parameters showed an increase during high light conditions. Suggesting its possible role in the protection of photosynthesis process when available light becomes intolerable to the host plants.

6.5 ROS Regulation and Control

There exist many reactions in aerobic metabolism that as a biproduct produce ROS in both plants as well as animals. These ROS produced under standard conditions can easily be detoxified by the natural defence mechanisms present in the organisms. However, under adverse environmental conditions these reactive species become a problem because many extreme conditions can significantly increase the production of ROS which is a primary cause for the damage caused by stresses to the cells and metabolism of the plants. ROS produced in the plants consist of both radical and non-radical species with peroxide, superoxide, and hydroxyl free radical being the most significant ROS in the plant system, with chloroplasts and mitochondria being the primary sites for their production owing to the high energy reactions taking place in these compartments. One of the major causes of ROS production in chloroplasts is the interaction between light and chlorophyll (Hasanuzzaman et al. 2020). The two important processes leading to the production of ROS in chloroplasts are related to photosystem I (PSI) and photosystem II (PSII) reactions. The capture of light energy at PSII can sometimes lead to the production of singlet oxygen by transferring the absorbed light energy to ground state oxygen molecule (Li and Kim 2021); whereas, at PSI superoxide radical is produced by Mehler's reaction when it absorbs light energy that can be further converted to peroxide or hydroxyl radical enzymatically (Hasanuzzaman et al. 2020).

It has been reported time and again that any disbalance between oxidant-antioxidant equilibrium is the major cause for the damage to plant tissues during environmental stresses (Nguyen et al. 2018). For example, singlet oxygen produced at PSII can damage major proteins of PSII and many amino acid residues of important photosynthetic proteins (Li and Kim 2021). Another important determinant of ROS damage is the interaction of intracellular ROS with signalling molecule nitrous oxide (NO) that can lead to the generation of reactive nitrogen species (RNS) when plants experience unfavourable conditions leading to a condition called as nitrosative stress (Corpas and Barroso 2013). Both these species, viz. ROS and RNS act as a signalling determinant at low concentrations while, at high concentrations they show toxic effects on cells as well as proteins. In the array of molecules used as antioxidants by plants, melatonin holds its importance because it is regulated by ROS and can also aid in detoxification of ROS. It can directly interact with ROS and regulate plant defences by scavenging the ROS produced during different reactions (Fig. 6.2). Some of the examples for involvement of melatonin in ROS regulation during stress are presented in the preceding section.

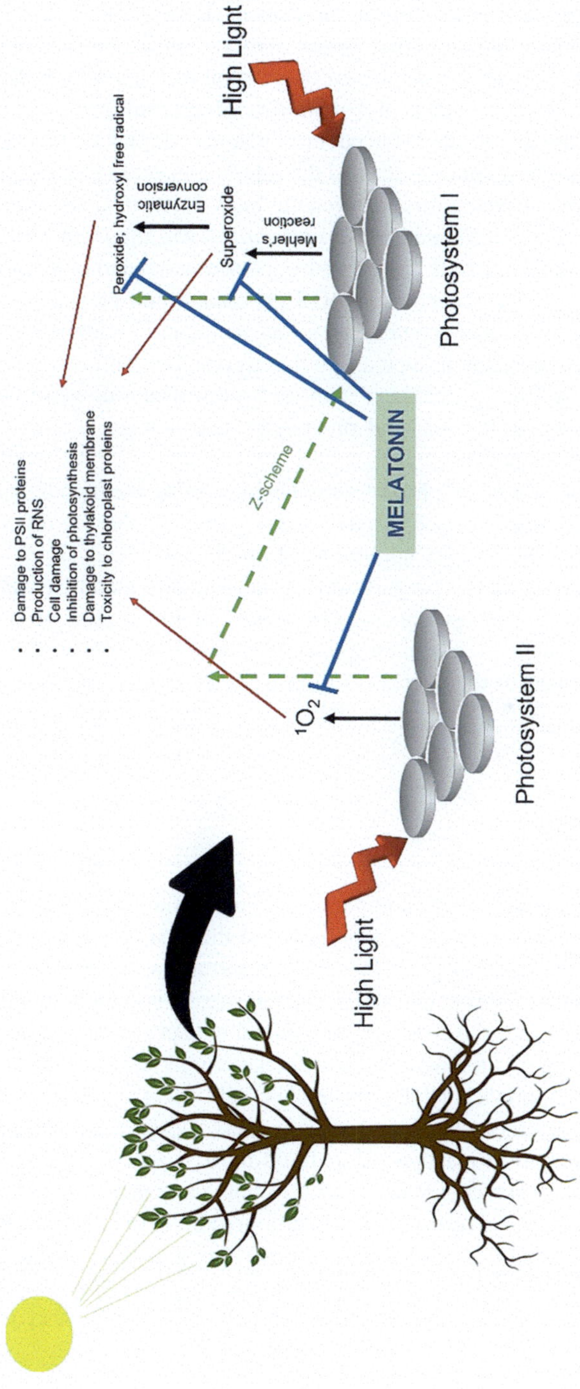

Fig. 6.2 Scheme for the generation of ROS in photosynthetic electron transfer in PSI and PSII and the possible role of melatonin in their scavenging

In an investigation conducted on *Arabidopsis* it was observed that induction of high light stress can lead to increase in the content of melatonin in the plants. This was further confirmed by analysing genes involved in its biosynthetic pathway including *SNAT1*. The authors confirmed the role on melatonin in high light stress by analysing *snat1* and *flu* mutants of *Arabidopsis* which confirmed the chloroplastic induction of melatonin which behaved in a manner similar to the exogenously applied melatonin. The authors confirmed the role of melatonin in protection against high light stress by analysing different genetic lines of *Arabidopsis* (Lee and Back 2018). In another study on *Haematococcus pluvialis* it was shown that melatonin can improve the resistance of plants towards nitrogen starvation and high light stress. In the study authors proved reduction in ROS generation and prevention of the cell damage in response to stress conditions, when melatonin is applied to the plants exogenously. They reported an increase in the astaxanthin (antioxidant compound) in the plant cells following treatment with melatonin which proved to be a major player in controlling ROS burst and preventing cell damage in stress conditions (Ding et al. 2018).

One of the major causes of photosystem damage during the high light conditions is the production of reactive species such as H_2O_2 and O_2^- that are produced in the chloroplasts during high light conditions. Experimental analysis of the plants exposed to high light show a significant increase in the mentioned ROS which further results in chloroplast damage. However, in the plants pre-treated with melatonin the damage to the chloroplasts was not observed as the melatonin pre-treatment resulted in the decrease in the ROS generation (Sharma and Zheng 2019; Sharma et al. 2020; Yang et al. 2021).

6.6 Conclusion

Melatonin is a ubiquitous indoleamine molecule present in the plants that functions under standard as well as stressed conditions. In the past two decades the role of melatonin in maintenance of plant metabolism and allowing the plant to succeed reproductively when experiencing environmental stress has been shown under many different conditions since the discovery of melatonin. Especially under the high light conditions focussing on the involvement of melatonin in maintenance of photosynthesis is of prime importance. High light and UV light affect plants in two ways; they can damage photosynthetic machinery such as photosystems and chlorophyll or it can lead to production of reactive oxygen and nitrosyl species and melatonin has been demonstrated to prevent the occurrence of both these damaging events. Hence, we conclude that focus on research specifically pertaining to effect of high light on plants and role of melatonin in photoprotection is required especially in the today's scenario when we are experiencing major climate changes. Also, considering the climate change we need to redefine what we call as high light that will lead to damage to the plants.

References

Arnao MB, Hernández-Ruiz J (2007) Melatonin promotes adventitious and lateral root regeneration in etiolated hypocotyls of *Lupinus albus* L. J Pineal Res 42:147–152

Arnao MB, Hernández-Ruiz J (2013) Growth conditions influence the melatonin content of tomato plants. Food Chem 138:1212–1214

Byeon Y, Back K (2015) Molecular cloning of melatonin 2-hydroxylase responsible for 2-hydroxymelatonin production in rice (*Oryza sativa*). J Pineal Res 58:343–351. https://doi.org/10.1111/jpi.12220

Conti A, Tettamanti C, Singaravel M, Haldar C, Pandi-Perumal SR, Maestroni GJM (2002) Melatonin: an ubiquitous and evolutionary hormone. In: Haldar C, Singaravel M, Maitra SK (eds) Treatise on pineal gland and melatonin. Science Publishers, Enfield, pp 105–143

Corpas FJ, Barroso JB (2013) Nitro-oxidative stress vs oxidative or nitrosative stress in higher plants. New Phytol 199:633–635. https://doi.org/10.1111/nph.12380

Ding F, Wang G, Wang M, Zhang S (2018) Exogenous melatonin improves tolerance to water deficit by promoting cuticle formation in tomato plants. Molecules 23:1605. https://doi.org/10.3390/molecules23071605

Dubbels R, Reiter R, Klenke E, Goebel A, Schnakenberg E, Ehlers C (1995) Melatonin in edible plants identified by radioimmunoassay and by high performance liquid chromatography-mass spectrometry. J Pineal Res 18:28–31. https://doi.org/10.1111/j.1600-079X.1995.tb00136.x

Hasanuzzaman M, Bhuyan MHMB, Parvin K, Bhuiyan TF, Anee TI, Nahar K (2020) Regulation of ROS metabolism in plants under environmental stress: a review of recent experimental evidence. Int J Mol Sci 21:8695. https://doi.org/10.3390/ijms21228695

Hattori A, Migitaka H, Iigo M, Itoh M, Yamamoto K, Ohtani-Kaneko R (1995) Identification of melatonin in plants and its effects on plasma melatonin levels and binding to melatonin receptors in vertebrates. Biochem Mol Biol Int 35:627–634

Hernández-Ruiz J, Arnao MB (2008) Distribution of melatonin in different zones of lupin and barley plants at different ages in the presence and absence of light. J Agric Food Chem 56: 10567–10573

Kołodziejczyk I, Bałabusta M, Szewczyk R, Posmyk MMJAPP (2015) The levels of melatonin and its metabolites in conditioned corn (*Zea mays* L.) and cucumber (*Cucumis sativus* L.) seeds during storage. Acta Physiol Plant 37:105. https://doi.org/10.1007/s11738-015-1850-7

Lee HY, Back K (2018) Melatonin induction and its role in high light stress tolerance in Arabidopsis thaliana. J Pineal Res 65:e12504. https://doi.org/10.1111/jpi.12504

Lee K, Zawadzka A, Czarnocki Z, Reiter RJ, Back K (2016) Molecular cloning of melatonin 3-hydroxylase and its production of cyclic 3-hydroxymelatonin in rice (*Oryza sativa*). J Pineal Res 61:470–478. https://doi.org/10.1111/jpi.12361

Li M, Kim C (2021) Chloroplast ROS and stress signaling. Plant Commun 3:100264. https://doi.org/10.1016/j.xplc.2021.100264

Manchester LC, Coto-Montes A, Boga JA, Andersen LPH, Zhou Z, Galano A (2015) Melatonin: an ancient molecule that makes oxygen metabolically tolerable. J Pineal Res 59:403–419. https://doi.org/10.1111/jpi.12267

Nazir M, Asad Ullah M, Mumtaz S, Siddiquah A, Shah M, Drouet S (2020) Interactive effect of melatonin and UV-C on phenylpropanoid metabolite production and antioxidant potential in callus cultures of purple basil (*Ocimum basilicum* L. var. purpurascens). Molecules 25:1072. https://doi.org/10.3390/molecules25051072

Nguyen H, Lin K, Ho S, Chiang C, Yang C (2018) Enhancing the abiotic stress tolerance of plants: from chemical treatment to biotechnological approaches. Physiol Plant 164:452–466. https://doi.org/10.1111/ppl.12812

Riga P, Medina S, Garcia-Flores LA, Gil-Izquierdo A (2014) Melatonin content of pepper and tomato fruits: effects of cultivar and solar radiation. Food Chem 156:347–352

Sarrou E, Therios I, Dimassi-Theriou K (2014) Melatonin and other factors that promote rooting and sprouting of shoot cuttings in *Punica granatum* cv. wonderful. Turkish J Bot 38:293–301

Sharma A, Zheng B (2019) Melatonin mediated regulation of drought stress: physiological and molecular aspects. Plan Theory 8(7):190

Sharma A, Wang J, Xu D, Tao S, Chong S, Yan D et al (2020) Melatonin regulates the functional components of photosynthesis, antioxidant system, gene expression, and metabolic pathways to induce drought resistance in grafted Carya cathayensis plants. Sci Total Environ 713:136675

Simopoulos A, Tan DX, Manchester LC, Reiter RJ (2005) Purslane: a plant source of omega-3 fatty acids and melatonin. J Pineal Res 39:331–332

Sun C, Liu L, Wang L, Li B, Jin C, Lin X (2021) Melatonin: a master regulator of plant development and stress responses. J Integr Plant Biol 63:126–145. https://doi.org/10.1111/jipb.12993

Tan DX, Reiter RJ (2020) An evolutionary view of melatonin synthesis and metabolism related to its biological functions in plants. J Exp Bot 71:4677–4689. https://doi.org/10.1093/jxb/eraa235

Tan DX, Manchester LC, Di Mascio P, Martinez GR, Prado FM, Reiter RJ (2007) Novel rhythms of N1-acetyl-N2-formyl-5-methoxykynuramine and its precursor melatonin in water hyacinth: importance for phytoremediation. FASEB J 21:1724–1729

Wu Y, Fan X, Zhang Y, Jiang J, Sun L, Rahman FU (2021) VvSNAT1 overexpression enhances melatonin production and salt tolerance in transgenic Arabidopsis. Plant Physiol Biochem 166: 485–494. https://doi.org/10.1016/j.plaphy.2021.06.025

Yang SJ, Huang B, Zhao YQ, Hu D, Chen T, Ding CB (2021) Melatonin enhanced the tolerance of Arabidopsis thaliana to high light through improving anti-oxidative system and photosynthesis. Front Plant Sci 12:752584. https://doi.org/10.3389/fpls.2021.752584

Ye T, Yin X, Yu L, Zheng SJ, Cai WJ, Wu Y (2019) Metabolic analysis of the melatonin biosynthesis pathway using chemical labeling coupled with liquid chromatography-mass spectrometry. J Pineal Res 66:e12531. https://doi.org/10.1111/jpi.12531

Zeng W, Mostafa S, Lu Z, Jin B (2022) Melatonin mediated abiotic stress tolerance in plants. Front Plant Sci 13:847175. https://doi.org/10.3389/fpls.2022.847175

Zhang N, Zhang HJ, Zhao B (2014) The RNA-seq approach to discriminate gene expression profiles in response to melatonin on cucumber lateral root formation. J Pineal Res 56:39–50

Zhang N, Sun Q, Zhang H, Cao Y, Weeda S, Ren S (2015) Roles of melatonin in abiotic stress resistance in plants. J Exp Bot 66:647–656. https://doi.org/10.1093/jxb/eru336

Zhang H, Wang L, Shi K, Shan D, Zhu Y, Wang C (2019) Apple tree flowering is mediated by low level of melatonin under the regulation of seasonal light signal. J Pineal Res 66:e12551. https://doi.org/10.1111/jpi.12551

Chapter 7
Melatonin in Plant Tolerance to Drought Stress: Physiological and Molecular Interventions

Harleen Kaur, Ravneet Kaur, and Ashish Sharma

Abstract Plants are subjected to various abiotic stresses; one such abiotic stress of great concern is drought stress. Drought stress affects plant metabolism and functioning of plants decreasing overall plant productivity. Plants have innate ability to tolerate harsh environmental conditions. Various signaling molecules are produced in plants in response to any abiotic stress. One such signaling molecule is melatonin. Melatonin is known to have a protective role in plants, it functions as an antioxidant that eliminates or reduces the levels of reactive oxygen species (ROS) produced under drought stress thus restoring the damage caused by drought stress in plants. Melatonin also increases the level of antioxidant enzymes in plants which further provide tolerance to plants against water-deficit conditions. Endogenous and exogenous melatonin can activate various stress-responsive genes, cascade of signaling pathways which further upregulate defense-related pathways and downregulate the pathways that are activated by drought stress.

Keywords Abiotic stress · Drought stress · Melatonin · Plant stress physiology · Phytomelatonin

7.1 Drought Stress

Plants are subjected to various stresses, which are broadly classified into two categories: biotic stress and abiotic stress. Both of these stress types have negative impact on overall plant development, ultimately responsible for reduced plant productivity. Various environmental catastrophe responsible for causing abiotic stress in plants are water deficiency, high/low temperatures, ultraviolet light, excessive salt, and heavy metals (Hoque et al. 2021). It is estimated that worldwide, abiotic stress causes 73% decline in crop production of cotton thereby indicating the threat caused by abiotic stresses to plant productivity (Mahmood et al. 2019).

H. Kaur · R. Kaur · A. Sharma (✉)
Department of Botany and Environment Science, DAV University, Jalandhar, Punjab, India
e-mail: ashish10210@davuniversity.org

A. Sharma, G. J. Ahammed (eds.), *Melatonin in Plants: Role in Plant Growth, Development, and Stress Response*, Plant Life and Environment Dynamics,
https://doi.org/10.1007/978-981-99-8051-2_7

Abiotic stresses are manifested as osmotic stress, oxidative stress, ionic imbalance, and cell metabolism dyshomeostasis, all of which collectively damages plant machinery thus affecting overall plant development (Hoque et al. 2021).

Amongst various abiotic stresses faced by plants, one of the significant abiotic stresses is drought stress. Water is found to have an important role in plant functioning, drought refers to low availability of water for prolonged time thereby affecting plant functioning (Mahmood et al. 2019), causing decline in crop productivity and ultimately causing serious damage to agricultural productivity (Yang et al. 2021). It has been observed that damage caused by drought stress to crop productivity worldwide is equivalent to the damage caused by all other abiotic stresses collectively (Manivannan et al. 2008). Apart from affecting plant development, drought stress is known to disturb various other plant functions such as plant respiration, photosynthesis, and stomatal movement (Yang et al. 2021), carbon assimilation rate, reduced turgor pressure, increased oxidative damage, and changes in leaf gas exchange (Nadeem et al. 2019) thereby, affecting plant growth and physiological metabolism.

Low availability of water to plants resulted in excessive production of reactive oxygen species (ROS) including O_2, O^{2-}, and H_2O_2, owing to their highly reactive nature, they cause damage to cell membranes, living tissues and macromolecules such as DNA, lipids, protein, carbohydrates ultimately causing programmed cell death (PCD) (Khaleghi et al. 2019). In order to reduce the damage caused by ROS, plant have evolved several cellular and molecular signaling pathways to activate and regulate defense mechanisms against biotic and abiotic stresses (Mahmood et al. 2019). One such defense mechanism is the activation of antioxidant machinery of plants. Various antioxidant enzymes, viz. APX, CAT, DHAR, GR, GST, MDHAR, POD, and SOD and non-enzymatic antioxidants (AsA, DHA, GSH, and GSSG) molecules are activated in response to excessive production of ROS that effectively reduce the ROS accumulation, thereby enabling plants to cope with stress, hence conferring tolerance against drought stress (Altaf et al. 2022).

Various studies in recent years revealed a new signaling molecule in plants termed melatonin that plays a vital role in protecting plants from harsh environmental conditions and exogenous application of melatonin has been shown to have a regulatory role in plants by restoring the losses caused by abiotic stress on plants (Pardo-Hernández et al. 2020). Previous studies revealed that almost all stress factors including drought stress, pH, cold stress as well as chemical stress agents such as $ZnSO_4$, NaCl, and H_2O_2, upregulated the expression of melatonin biosynthesis genes thereby increasing the level of melatonin in response to any kind of stress in plants (Arnao and Hernández-Ruiz 2013). The elevated level of melatonin in response to any harsh environmental condition may serve as an adaptive response of plants to tolerate adverse environmental conditions owing to its antioxidant properties (Franzoni et al. 2019).

7.2 Melatonin in Plants: A Versatile Antioxidant

Melatonin (N-acetyl-5-methoxytrytamine, MT) is a familiar pleiotropic signaling molecule that is known to function as an antioxidant in plants thereby conferring tolerance to plants against abiotic and biotic stresses (Hoque et al. 2021). It is an indolic compound (biogenic indoleamine) whose structure is related to other important substances present in plants such as tryptophan, serotonin, indole-3-acetic acid (IAA), etc. (Arnao and Hernández-Ruiz 2006).

Melatonin occurs ubiquitously in nature and its actions are thought to represent one of the most phylogenetically ancient of all biological signaling mechanisms (Pandi-Perumal et al. 2006). Its presence can be traced to primitive photosynthetic bacteria, red and green algae, fungi, and plants (Tan 2015). Melatonin was first discovered in plants in 1995 (Dubbels et al. 1995; Hattori et al. 1995). In plants, the synthesis of melatonin from tryptophan (an amino acid) was first reported around 20 years back. Metabolism of melatonin resulted in the production of various important bioactive compounds in plants generally grouped as: (a) products of oxidation reactions, (b) products of catabolism, and (c) conjugates and derivatives (Murch and Erland 2021). Plant melatonin, also referred to as "phytomelatonin" (Blask et al. 2004), was studied as a naturally occurring antioxidants as well as functions as a plants growth regulator by regulating the process of rooting, growth, modulator of gene expression, and various other morphogenetic features (Arnao and Hernández-Ruiz 2015).

One of the functions of plant growth regulators in addition to performing function of regulating plant development is to providing tolerance to plants against various abiotic stresses. Significant production and levels of melatonin have been found in majority of monocotyledons and dicotyledons families (Rehaman et al. 2021). In dry seeds or dormant seeds, high level of melatonin function as antioxidants and were found to be responsible for conferring tolerance as in such cases enzymes are less effective and cannot be upregulated (Pandi-Perumal et al. 2006).

Melatonin perform numerous functions in cell such as a free radical scavenger as well as a powerful antioxidant (Arnao and Hernández-Ruiz 2015). In response to stress, melatonin act as an antioxidant that helps plant to tolerate harsh conditions as well as provides protection to photosynthetic machinery which otherwise gets damaged and helps to protect germ tissues. The capacity to neutralize ROS also plays an important role in melatonin's control of ROS and reactive nitrogen species (RNS) signaling networks within plants, in particular hydrogen peroxide (H_2O_2) and nitric oxide (NO) signaling networks which allow for rapid signaling in response to stimulus (Lee and Back 2017). Mechanistically, it has been observed that one molecule of melatonin eliminated around 10 units of ROS without any requirement of recycling process (Rodriguez-Naranjo et al. 2012).

In stressed environments, various harmful reactive molecules occur in plant cells such as hydroxyl radical, superoxide anion, singlet oxygen, hydrogen peroxide, hypochlorous acid, nitric oxide, peroxynitrite anion, peroxynitrous acid, and lipid peroxyl radical, among others. Melatonin function to eliminate these hazardous molecules by acting as a free radical scavenger. In addition to this, melatonin functions as a regulator of many redox enzymes in animal tissues, increasing the expression of antioxidant enzymes such as catalases, peroxidases, and superoxide dismutases, and decreasing the expression of pro-oxidant enzymes (Arnao and Hernández-Ruiz 2015).

Although melatonin function as an antioxidant in plants, however, melatonin molecule does not display any pro-oxidative effects, while melatonin-intermediate products show antioxidant properties and function synergistically with other antioxidants in plants such as ascorbic acid, glutathione, etc. (Arnao and Hernández-Ruiz 2006).

7.3 Role of Melatonin in Plants Subjected to Drought Stress: Physiological Responses

In response to drought stress, various adaptive strategies are adopted by plants that help plant to balance excessive levels of ROS being produced and in turn protecting the plants form drought induced harmful effects. However, these adaptive responses are functional up to a certain level. Prolonged exposure of severe abiotic stresses obstruct these inbuilt defense systems of plants thereby affecting plant functioning, the outcomes of which appear as reduced photosynthesis, stomatal conductance, leaf area, growth rate, plant height, and water potential (Tiwari et al. 2022). Various stress alleviating substances are known to enhance plant tolerance against abiotic stress such as melatonin which is found to be beneficial for plants exposed to abiotic stresses such as drought, high/low temperature, salinity, heavy metals, chemicals, and pathogens (Kaya et al. 2020). Figure 7.1 displays some of the negative effect of drought stress on plants as well as role of melatonin in counteracting such effects in drought stressed plants.

One of the most significant ways by which melatonin provides tolerance to drought stressed plants is the enhancement of enzymes involved in antioxidant defense system and ROS scavenging (Sharma et al. 2020). Superoxide anions produced in response to drought stress are eliminated from plant cell by the action of melatonin either via enhanced scavenging activity or by reduced production of superoxide anions. Furthermore, melatonin also improves scavenging proficiency of H_2O_2 in plants growing under water deficient environment (Sharma and Zheng 2019). The detrimental effect of drought stress on seed germination can be reversed by activity of melatonin (Zhang et al. 2013). Under drought stress, melatonin causes inhibition of membrane peroxidation and protein denaturation thereby resulting in enhancement in germination rate.

Exogenous application of melatonin helps plant to withstand drought conditions by improving plant physiological characters thereby improving overall plant development (Tiwari et al. 2021). In drought stressed environment, melatonin maintains homeostasis balance of growth of vegetative tissues thereby conferring tolerance to

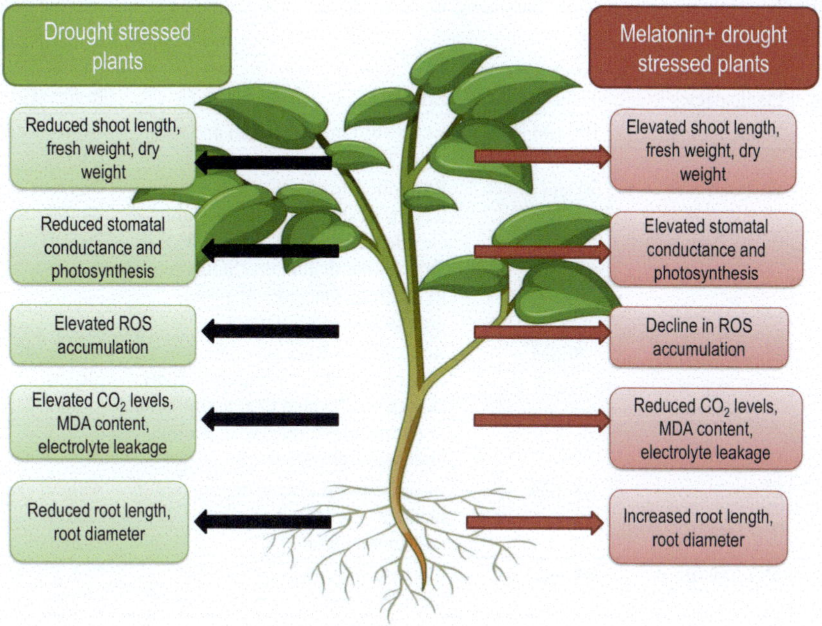

Fig. 7.1 Beneficial role of melatonin in plants subjected to drought stress. Plants subjected to drought stress experience reduced physiological activity which can be alleviated by the application of melatonin to plants

plants (Sharma and Zheng 2019). Under drought stress, melatonin protects the photosynthetic pigment (chlorophyll) from degradation. Not only principle photosynthetic pigments, melatonin also protects accessory photosynthetic pigments such as carotenoids from degradation, thereby improving the photosynthetic efficiency.

Melatonin plays a vital role in the promotion of seminal root growth in rice seedling. The seminal root growth without alteration of adventitious root number in transgenic rice seedling suggests that melatonin as a hormone is important for root development in seedlings (Park and Back 2012). Melatonin resembles plant growth hormone, auxin in terms of physiological activity performed by both of them in plants such as growth promotion of seedlings under drought stress (Li et al. 2018). In addition to above-mentioned functions of melatonin, various other beneficial effects of exogenous application of melatonin in plant growing under drought stress are enlisted in Table 7.1.

Table 7.1 Role of melatonin in alleviating negative effects of drought stress in different plant species

S. No.	Plant species	Concentration of melatonin	Effect	Reference
1	Maize	100 μM	Expansion in root length and root diameter	Ahmad et al. (2019)
2	Soybean	50 and 100 μM	Promotion in shoot length, shoot length, fresh weight, dry weight, elevated Catalase (CAT), and Superoxide dismutase (SOD) activity, thereby causing enhanced growth and development	Imran et al. (2021a, 2021b)
3	Maize seedlings	100 μM	Reduction in ROS accumulation and enhanced ROS scavenging activity as well as elevated CO_2 assimilation in plants which otherwise decline under drought stress	Huang et al. (2019)
4	Maize	100 μM	Increase in relative water content in plants as compared to plants non-treated plants	Ahmad et al. (2021a, 2021b)
5	*M. oleifera*	100 mM	Increase in growth parameters including plant height, number of branches/plant fresh, and dry weights of leaves and shoots as well as enhanced indole acetic acid and phenolic contents	Sadak et al. (2020)
6	Moldavian balm (*Dracocephalum moldavica*)	100 μM	An increment in polyphenol compounds as well as increased protein content	Naghizadeh et al. (2019)
7	Coffee (*Coffea arabica*)		Inhibition of the increase of both MDA content and electrolyte leakage (EL) of coffee plants under drought stress	Cherono et al. (2021)
8	Cucumber (*Cucumis sativus* L.)	500 μM	Increased volume of whole seedlings, increased root:shoot ratio, increased stomatal conductance and net photosynthesis in plants	Zhang et al. (2013)
9	Tomato		Increased root length, root volume, root surface area, root forks and projected area. Enzymatic activities, viz. SOD, CAT, APX, GR, and POD were enhanced	Altaf et al. (2022)
10	Kiwifruit (*A. chinensis* var. deliciosa cv. Qinmei)	100 μM	Elevation in proline and soluble protein contents	Liang et al. (2019)
11	Tomato	0.2 mM	Reduction in MDA and superoxide anions as compared to plants under drought stress, also caused reduction in Ascorbate content	Liu et al. (2015)

(continued)

Table 7.1 (continued)

S. No.	Plant species	Concentration of melatonin	Effect	Reference
12	Wheat (*Triticum aestivum* L. "Yan 995")	500 μM	Increased content of glutathione and ascorbate in leaves	Cui et al. (2017)
13	Naked oat (*Avena nuda* L.)	100 μM	Promotion in the growth of seedlings as well as increase in antioxidant enzyme activities including SOD, POD, CAT, and APX	Gao et al. (2018)
14	Citrus	100 μM	Increase in the length of seedlings	Jafari and Shahsavar (2020)
15	Apple (*Malus domestica* Borkh)	100 μM	Small increase in CO_2 concentration as compared to plants grown under drought stress	Wang et al. (2013)
16	*Medicago sativa*		Melatonin-treated plants displayed lower NR activity as compared to non-treated stressed plants having higher NR activity	Antoniou et al. (2017)
17	*Zea mays* L.	100 μM	Melatonin causes reduction in the amount of H_2O_2 accumulated in plants under drought stress	Ye et al. (2016)
18	Rice seedlings (*Oryza sativa* L.)	200 μM	Melatonin pretreatment causes increased levels of antioxidant enzymes, viz. SOD, POD, and CAT in leaves	Luo et al. (2022)
19	Micro-Tom tomato	100 μM	Melatonin treatment caused elevated relative water content (RWC)	Mushtaq et al. (2022)
20	Fenugreek (*Trigonella foenum-gracum* L.)	100 and 300 μM	Melatonin caused elevated fresh weight of shoot	Zamani et al. (2020)
21	*Cucumis sativus* L.	100 μM	Protection form drought stressed conditions dure to elevated antioxidant enzyme activities in plant	Zhang et al. (2013)
22	*Dracocephalum moldavica*	100 μM	Protection from drought stress by enhanced proline content, chlorophyll content, and relative water content	Kabiri et al. (2018)
23	*Actinidia chinensis*	50 and 100 μM	Induced ascorbic acid-glutathione (AsA-GSH) cycle, carotenoid biosynthesis, and increased defense enzymes activities resulting in improve seedling growth under drought stress	Xia et al. (2020)
24	*Cicer arietinum* L.	0.25, 0.50 and 1.0 mM	Increased chlorophyll a, b and carotenoids content in plant treated with melatonin	Dawood (2017)

(continued)

Table 7.1 (continued)

S. No.	Plant species	Concentration of melatonin	Effect	Reference
25	*Agrostis stolonifera*	20 μM	Electrolyte leakage and MDA content in ME-treated ipt-transgenic plants were lower than those of the untreated ipt-transgenic plants	Ma et al. (2018)

7.4 Melatonin-Mediated Tolerance to Drought Stresses Plants: Molecular Mechanism

Abiotic stress tolerance mediated by melatonin seems to be regulated at various levels including receptor binding and expression, regulation of stress-responsive genes at transcriptional level, regulation of various signaling pathways governed by calcium-dependent kinase, and accumulation of compatible solutes. When a plant suffers any abiotic stress such as high/low temperature, heavy metal stress, UV radiations, drought stress and waterlogging, expression of biosynthesis enzyme transcripts such as TDC, SNAT, ASMT, and COMT genes increases, leading to elevated level of endogenous melatonin that function to protect plants from damage caused by abiotic stress (Ahmad et al. 2021a, b).

Elevated level of melatonin causes further expression of various factors stress receptors, kinases, and transcription factors which are involved in plant defense toward abiotic stress (Weeda et al. 2014). Various transcription factors that function to upregulate expression of stress-responsive genes are governed by melatonin treatment. Apart from regulation mediated by endogenous melatonin, various transcriptional factors such as BASIC LEUCINE ZIPPER PROTEINS, CBF/DREBs, MYB, and zinc finger protein are upregulated in response to exogenous application of melatonin to plants growing under abiotic stress. It has been found that various stress-responsive genes such as C-repeat-binding factors (CBFs), drought-responsive element-binding (DREB) factors, and downstream cold-responsive genes, such as COR15a were regulated by melatonin whereas certain transcriptional factors such as CAMTA1, ZAT6, ZAT10, and AZT12 involved in freezing and drought stress tolerance, were activated by melatonin (Li et al. 2015). Previous study revealed that mitogen-associated protein kinases (MAPKs) and transcription factors (TF) are activated by melatonin, also regulate expression of downstream stress-responsive genes, thus helping plants tolerate drought stress (Gao et al. 2018). When subjected to abiotic stress, the DREB transcription factors are activated by abiotic stresses to regulate stress signal transduction and gene expression. Various transcription factors such as bZIP and DREB2 help plant conferring tolerance against drought stress and plant growth (Imran et al. 2021a, b).

Melatonin upregulates the expression of genes involved in photosynthesis, carbohydrate accumulation, and phloem loading of sucrose (Zhao et al. 2015). It has been found that melatonin upregulates expression of PsaK and PsaG, which in turn causes further growth of soybean plants. PsbO (oxygen-evolving enhancer protein

1/OEE1) and PsbP (OEE2) play important role in plant. Optimum expression of these two influences activity of OEC thus regulating plant growth (Wei et al. 2015). PsbO (oxygen-evolving enhancer protein 1/OEE1) is necessary for the stabilization of the cluster; and PsbP (OEE2) is required for the oxygen-evolving activity.

Under drought stress, plant's chlorophyll content reduces which is due to activity of certain chlorophyll degrading enzymes, viz. chlorophyllase (CHLASE), pheophytinase (PPH), and pheophorbidea-oxygenase (PAO), the activity of which is increased due to upregulation of genes encoding such enzymes under drought. Melatonin helps in restoring the chlorophyll content in plants probably due to downregulation of expression of CHLASE, PPH, and PAO in plants exposed to water-deficit stress (Ma et al. 2018; Wang et al. 2013). Furthermore, melatonin upregulates the expression of gene PetF ferredoxin which is involved in regulating the amount of reduced ascorbate thereby preventing chlorophyll degradation. Moreover, melatonin upregulates the ATP transporter-related genes in plants under stressful conditions (Gu et al. 2017; Hasan et al. 2015).

As an effect of drought stress, ROS/RNS level increases in plants, however, melatonin controls the levels of ROS/RNS by elevating the activity of antioxidant-related genes, viz. Superoxide dismutase (SOD), Peroxidase (POD), Catalse (CAT), Ascorbate peroxidase (APX), glutathione S-transferase (GST), monodehydroascorbate reductase (MDHAR), dehydroascorbate reductase (DHAR), and glutathione reductase (GR), also melatonin upregulates the expression of proline biosynthesis genes thereby regulates the activity of osmoprotectants in plants subjected to drought stress (Moustafa-Farag et al. 2020). This melatonin-mediated regulation at the gene level can be attributed to the modulation of key cell signaling pathways such as the mitogen-activated protein kinase (MAPK) cascade pathway, which is also regulated by melatonin under drought conditions (Sharma et al. 2020). Under drought stress, Melatonin-mediated activation of antioxidant machinery is linked to activation of MAPK cascade including activation of Asmap1 and Aspk11 as well as activation of WRKY1, DREB2, and MYB transcription factors, which further leads to expression of antioxidant-related genes (Tiwari et al. 2021).

Drought stress affects levels of plant hormones, it increases the levels of ABA, BRs, and JA while decreases the level of CKs and GAs, while melatonin is found to have opposing effect on plant hormones in plant under drought stress, i.e., it enhances the levels of BRs, GAs, JA, and CKs and decreases the level of ABA (Moustafa-Farag et al. 2020). Decline in ABA level in response to exogenous application of melatonin to drought stressed plants is due to the upregulation of ABA degradation gene, viz. MdCYP707A1 and MDCYP707A1 as well as downregulation of ABA biosynthetic gene, viz. 9-cis-epoxycarotenoid dioxygenase 3 and NCED3 (Li et al. 2015). Moreover, melatonin regulates abscisic acid signaling-related genes such as SnRK2 (SNF1-related protein kinases 2), RCAR/PYR/PYL, and NCED (nine-cis-epoxycarotenoid dioxygenase). Increased level of ABA in drought stressed plants, negatively affects plant metabolic pathways including closing of stomata, such changes thereby contributing to leaf aging (Moustafa-Farag et al. 2020). Exogenous application of melatonin in rhizosphere of drought stressed plant upregulates the mRNA level of TIP aquaporins (Kurowska et al. 2019), thereby increasing the water uptake capacity of plant hence decreasing water loss (Qiao et al. 2020).

7.5 Conclusion

Plants face various abiotic stresses and develop various adaptive strategies to cope with such stresses. Various signaling molecules perform varied function in plants and help in plant functioning. Melatonin is one of the signaling molecule that function as an antioxidant by conferring tolerance to plants under abiotic stressed environments. In response to abiotic stress particularly drought stress, melatonin activates various genes that further activates enzymes/molecules involved in providing tolerance to plants by reducing/restoring damage caused by plants. Exogenous application of melatonin has found to be beneficial in protecting plants from drought stress thereby increasing overall growth and productivity of plants.

References

Ahmad S, Kamran M, Ding R, Meng X, Wang H, Ahmad I et al (2019) Exogenous melatonin confers drought stress by promoting plant growth, photosynthetic capacity and antioxidant defense system of maize seedlings. PeerJ 7:e7793

Ahmad S, Cui W, Kamran M, Ahmad I, Meng X, Wu X et al (2021a) Exogenous application of melatonin induces tolerance to salt stress by improving the photosynthetic efficiency and antioxidant defense system of maize seedling. J Plant Growth Regul 40:1270–1283

Ahmad S, Muhammad I, Wang GY, Zeeshan M, Yang L, Ali I, Zhou XB (2021b) Ameliorative effect of melatonin improves drought tolerance by regulating growth, photosynthetic traits and leaf ultrastructure of maize seedlings. BMC Plant Biol 21(1):1–14

Altaf MA, Shahid R, Ren MX, Naz S, Altaf MM, Khan LU et al (2022) Melatonin improves drought stress tolerance of tomato by modulating plant growth, root architecture, photosynthesis, and antioxidant defense system. Antioxidants 11(2):309

Antoniou C, Chatzimichail G, Xenofontos R, Pavlou JJ, Panagiotou E, Christou A, Fotopoulos V (2017) Melatonin systemically ameliorates drought stress-induced damage in Medicago sativa plants by modulating nitro-oxidative homeostasis and proline metabolism. J Pineal Res 62(4): e12401

Arnao MB, Hernández-Ruiz J (2006) The physiological function of melatonin in plants. Plant Signal Behav 1(3):89–95

Arnao MB, Hernández-Ruiz J (2013) Growth conditions determine different melatonin levels in Lupinus albus L. J Pineal Res 55(2):149–155

Arnao MB, Hernández-Ruiz J (2015) Functions of melatonin in plants: a review. J Pineal Res 59(2): 133–150

Blask DE, Dauchy RT, Sauer LA, Krause JA (2004) Melatonin uptake and growth prevention in rat hepatoma 7288CTC in response to dietary melatonin: melatonin receptor-mediated inhibition of tumor linoleic acid metabolism to the growth signaling molecule 13-hydroxyoctadecadienoic acid and the potential role of phytomelatonin. Carcinogenesis 25(6):951–960

Cherono S, Ntini C, Wassie M, Mollah MD, Belal MA, Ogutu C, Han Y (2021) Exogenous application of melatonin improves drought tolerance in coffee by regulating photosynthetic efficiency and oxidative damage. J Am Soc Hortic Sci 146(1):24–32

Cui G, Zhao X, Liu S, Sun F, Zhang C, Xi Y (2017) Beneficial effects of melatonin in overcoming drought stress in wheat seedlings. Plant Physiol Biochem 118:138–149

Dawood MG (2017) Physiological effect of melatonin, IAA and their precursor on quality and quantity of chickpea plants grown under sandy soil conditions. Agric Eng Int CIGR J 19:35–44

Dubbels R, Reiter RJ, Klenke E, Goebel A, Schnakenberg E, Ehlers C et al (1995) Melatonin in edible plants identified by radioimmunoassay and by high performance liquid chromatography-mass spectrometry. J Pineal Res 18(1):28–31

Franzoni G, Trivellini A, Bulgari R, Cocetta G, Ferrante A (2019) Bioactive molecules as regulatory signals in plant responses to abiotic stresses. In: MIR K, Reddy PS, Ferrante A, Khan NA (eds) Plant signaling molecules. Woodhead Publishing, Cambridge, pp 169–182

Gao W, Zhang Y, Feng Z, Bai Q, He J, Wang Y (2018) Effects of melatonin on antioxidant capacity in naked oat seedlings under drought stress. Molecules 23(7):1580

Gu Q, Chen Z, Yu X, Cui W, Pan J, Zhao G et al (2017) Melatonin confers plant tolerance against cadmium stress via the decrease of cadmium accumulation and reestablishment of microRNA-mediated redox homeostasis. Plant Sci 261:28–37

Hasan MK, Ahammed GJ, Yin L, Shi K, Xia X, Zhou Y et al (2015) Melatonin mitigates cadmium phytotoxicity through modulation of phytochelatins biosynthesis, vacuolar sequestration, and antioxidant potential in Solanum lycopersicum L. Front Plant Sci 6:601

Hattori A, Migitaka H, Iigo M, Itoh M, Yamamoto K, Ohtani-Kaneko R et al (1995) Identification of melatonin in plants and its effects on plasma melatonin levels and binding to melatonin receptors in vertebrates. Biochem Mol Biol Int 35(3):627–634

Hoque MN, Tahjib-Ul-Arif M, Hannan A, Sultana N, Akhter S, Hasanuzzaman M et al (2021) Melatonin modulates plant tolerance to heavy metal stress: morphological responses to molecular mechanisms. Int J Mol Sci 22(21):11445

Huang B, Chen YE, Zhao YQ, Ding CB, Liao JQ, Hu C et al (2019) Exogenous melatonin alleviates oxidative damages and protects photosystem II in maize seedlings under drought stress. Front Plant Sci 10:677

Imran M, Latif Khan A, Shahzad R, Aaqil Khan M, Bilal S, Khan A et al (2021a) Exogenous melatonin induces drought stress tolerance by promoting plant growth and antioxidant defence system of soybean plants. AoB Plants 13(4):plab026

Imran M, Shazad R, Bilal S, Imran QM, Khan M, Kang SM et al (2021b) Exogenous melatonin mediates the regulation of endogenous nitric oxide in Glycine max L. to reduce effects of drought stress. Environ Exp Bot 188:104511

Jafari M, Shahsavar A (2020) The application of artificial neural networks in modeling and predicting the effects of melatonin on morphological responses of citrus to drought stress. PLoS One 15(10):e0240427

Kabiri R, Hatami A, Oloumi H, Naghizadeh M, Nasibi F, Tahmasebi Z (2018) Foliar application of melatonin induces tolerance to drought stress in Moldavian balm plants (Dracocephalum moldavica) through regulating the antioxidant system. Folia Hortic 30(1):155

Kaya C, Higgs D, Ashraf M, Alyemeni MN, Ahmad P (2020) Integrative roles of nitric oxide and hydrogen sulfide in melatonin-induced tolerance of pepper (Capsicum annuum L.) plants to iron deficiency and salt stress alone or in combination. Physiol Plant 168(2):256–277

Khaleghi A, Naderi R, Brunetti C, Maserti BE, Salami SA, Babalar M (2019) Morphological, physiochemical and antioxidant responses of Maclura pomifera to drought stress. Sci Rep 9(1):19250

Kurowska MM, Wiecha K, Gajek K, Szarejko I (2019) Drought stress and re-watering affect the abundance of TIP aquaporin transcripts in barley. PLoS One 14(12):e0226423

Lee HY, Back K (2017) Melatonin is required for H_2O_2-and NO-mediated defense signaling through MAPKKK 3 and OXI 1 in Arabidopsis thaliana. J Pineal Res 62(2):e12379

Li C, Tan DX, Liang D, Chang C, Jia D, Ma F (2015) Melatonin mediates the regulation of ABA metabolism, free-radical scavenging, and stomatal behaviour in two Malus species under drought stress. J Exp Bot 66(3):669–680

Li J, Zeng L, Cheng Y, Lu G, Fu G, Ma H et al (2018) Exogenous melatonin alleviates damage from drought stress in Brassica napus L. (rapeseed) seedlings. Acta Physiol Plant 40:1–11

Liang D, Ni Z, Xia H, Xie Y, Lv X, Wang J et al (2019) Exogenous melatonin promotes biomass accumulation and photosynthesis of kiwifruit seedlings under drought stress. Sci Hortic 246:34–43

Liu J, Wang W, Wang L, Sun Y (2015) Exogenous melatonin improves seedling health index and drought tolerance in tomato. Plant Growth Regul 77:317–326

Luo C, Min W, Akhtar M, Lu X, Bai X, Zhang Y et al (2022) Melatonin enhances drought tolerance in rice seedlings by modulating antioxidant systems, osmoregulation, and corresponding gene expression. Int J Mol Sci 23(20):12075

Ma X, Zhang J, Burgess P, Rossi S, Huang B (2018) Interactive effects of melatonin and cytokinin on alleviating drought-induced leaf senescence in creeping bentgrass (Agrostis stolonifera). Environ Exp Bot 145:1–11

Mahmood T, Khalid S, Abdullah M, Ahmed Z, Shah MKN, Ghafoor A, Du X (2019) Insights into drought stress signaling in plants and the molecular genetic basis of cotton drought tolerance. Cell 9(1):105

Manivannan P, Jaleel CA, Somasundaram R, Panneerselvam R (2008) Osmoregulation and antioxidant metabolism in drought-stressed Helianthus annuus under triadimefon drenching. C R Biol 331(6):418–425

Moustafa-Farag M, Mahmoud A, Arnao MB, Sheteiwy MS, Dafea M, Soltan M et al (2020) Melatonin-induced water stress tolerance in plants: recent advances. Antioxidants 9(9):809

Murch SJ, Erland LA (2021) A systematic review of melatonin in plants: an example of evolution of literature. Front Plant Sci 12:683047

Mushtaq N, Iqbal S, Hayat F, Raziq A, Ayaz A, Zaman W (2022) Melatonin in micro-tom tomato: improved drought tolerance via the regulation of the photosynthetic apparatus, membrane stability, osmoprotectants, and root system. Life 12(11):1922

Nadeem M, Li J, Yahya M, Sher A, Ma C, Wang X, Qiu L (2019) Research progress and perspective on drought stress in legumes: a review. Int J Mol Sci 20(10):2541

Naghizadeh M, Kabiri R, Hatami A, Oloumi H, Nasibi F, Tahmasei Z (2019) Exogenous application of melatonin mitigates the adverse effects of drought stress on morpho-physiological traits and secondary metabolites in Moldavian balm (Dracocephalum moldavica). Physiol Mol Biol Plants 25:881–894

Pandi-Perumal SR, Srinivasan V, Maestroni GJM, Cardinali DP, Poeggeler B, Hardeland R (2006) Melatonin: nature's most versatile biological signal? FEBS J 273(13):2813–2838

Pardo-Hernández M, López-Delacalle M, Rivero RM (2020) ROS and NO regulation by melatonin under abiotic stress in plants. Antioxidants 9(11):1078

Park S, Back K (2012) Melatonin promotes seminal root elongation and root growth in transgenic rice after germination. J Pineal Res 53(4):385–389

Qiao Y, Ren J, Yin L, Liu Y, Deng X, Liu P, Wang S (2020) Exogenous melatonin alleviates PEG-induced short-term water deficiency in maize by increasing hydraulic conductance. BMC Plant Biol 20:1–14

Rehaman A, Mishra AK, Ferdose A, Per TS, Hanief M, Jan AT, Asgher M (2021) Melatonin in plant defense against abiotic stress. Forests 12(10):1404

Rodriguez-Naranjo MI, Moyá ML, Cantos-Villar E, Garcia-Parrilla MC (2012) Comparative evaluation of the antioxidant activity of melatonin and related indoles. J Food Compos Anal 28(1):16–22

Sadak MS, Abdalla AM, Abd Elhamid EM, Ezzo MI (2020) Role of melatonin in improving growth, yield quantity and quality of Moringa oleifera L. plant under drought stress. Bullet Natl Res Centre 44(1):1–13

Sharma A, Zheng B (2019) Melatonin mediated regulation of drought stress: physiological and molecular aspects. Plan Theory 8(7):190

Sharma A, Wang J, Xu D, Tao S, Chong S, Yan D et al (2020) Melatonin regulates the functional components of photosynthesis, antioxidant system, gene expression, and metabolic pathways to induce drought resistance in grafted Carya cathayensis plants. Sci Total Environ 713:136675

Tan DX (2015) Melatonin and plants. J Exp Bot 66(3):625–626

Tiwari RK, Lal MK, Kumar R, Chourasia KN, Naga KC, Kumar D et al (2021) Mechanistic insights on melatonin-mediated drought stress mitigation in plants. Physiol Plant 172(2): 1212–1226

Tiwari RK, Lal MK, Kumar R, Mangal V, Altaf MA, Sharma S et al (2022) Insight into melatonin-mediated response and signaling in the regulation of plant defense under biotic stress. Plant Mol Biol 109:385–399

Wang P, Sun X, Li C, Wei Z, Liang D, Ma F (2013) Long-term exogenous application of melatonin delays drought-induced leaf senescence in apple. J Pineal Res 54(3):292–302

Weeda S, Zhang N, Zhao X, Ndip G, Guo Y, Buck GA et al (2014) Arabidopsis transcriptome analysis reveals key roles of melatonin in plant defense systems. PLoS One 9(3):e93462

Wei W, Li QT, Chu YN, Reiter RJ, Yu XM, Zhu DH et al (2015) Melatonin enhances plant growth and abiotic stress tolerance in soybean plants. J Exp Bot 66(3):695–707

Xia, H, Ni, Z, Hu, R, Lin, L, Deng, H, Wang, J, Liang, D. et al (2020). Melatonin alleviates drought stress by a non-enzymatic and enzymatic antioxidative system in kiwifruit seedlings. International Journal of Molecular Sciences, 21(3), 852.

Yang X, Lu M, Wang Y, Wang Y, Liu Z, Chen S (2021) Response mechanism of plants to drought stress. Horticulturae 7(3):50

Ye J, Wang S, Deng X, Yin L, Xiong B, Wang X (2016) Melatonin increased maize (Zea mays L.) seedling drought tolerance by alleviating drought-induced photosynthetic inhibition and oxidative damage. Acta Physiol Plant 38(2):48

Zamani Z, Amiri H, Ismaili A (2020) Improving drought stress tolerance in fenugreek (Trigonella foenum-graecum) by exogenous melatonin. Plant Biosyst 154(5):643–655

Zhang N, Zhao B, Zhang HJ, Weeda S, Yang C, Yang ZC et al (2013) Melatonin promotes water-stress tolerance, lateral root formation, and seed germination in cucumber (Cucumis sativus L.). J Pineal Res 54(1):15–23

Zhao H, Su T, Huo L, Wei H, Jiang Y, Xu L, Ma F (2015) Unveiling the mechanism of melatonin impacts on maize seedling growth: sugar metabolism as a case. J Pineal Res 59(2):255–266

Chapter 8
Effect of Melatonin in Regulating Salt Stress Responses in Plants

Vasudha Maurya, Narayan Singh, Indu Sharma, Ashutosh Sharma, and Rahul Kumar

Abstract Salt stress is one of the most significant constraints in global agricultural production by causing a significant yield loss, every year. Attempts have been made to mitigate the harmful effects of salt stress by using various bio-stimulants such as some biomolecules, plant extracts or soil-inhabiting microorganisms. One of such potential biomolecules is the melatonin that functions as potent growth regulator, bio-stimulant, antioxidant, and plant stress protectant. Melatonin is a multifunctional molecule found in both prokaryotes as well as eukaryotes and has received extensive validation for its ability to regulate plant stress tolerance, particularly its critical function in improving tolerance toward salt stress. The application of exogenous melatonin to the plant under salt stress has received a lot of attention from researchers as a stress ameliorative treatment. Various modes of application of exogenous melatonin have been tried and found effective in conferring a better tolerance toward salt stress. In order to boost antioxidant systems under salt stress, melatonin also plays crucial roles as an antioxidant and free radical scavenger. These processes enhance photosynthesis, maintain ion homeostasis, and trigger a number of downstream signals that regulate the production of other plant growth regulators. Attempts have also been made for increasing the endogenous level of melatonin by the over-expression of the genes involved in the biosynthetic pathway of melatonin, suggesting them as key gene candidates for enhancing the tolerance toward salt stress via the genetic transformation. Furthermore, melatonin also regulates some salt stress-responsive genes in plants. Overall, the present chapter presents melatonin as a potential bio-stimulant for mitigating the harmful effects of salt stress in plants,

V. Maurya · N. Singh
Department of Biotechnology, DAV University, Jalandhar, Punjab, India

Faculty of Agricultural Sciences, DAV University, Jalandhar, Punjab, India

I. Sharma
Department of Life Sciences, University Institute of Sciences, Sant Baba Bhag Singh University, Jalandhar, Punjab, India

A. Sharma (✉) · R. Kumar (✉)
Faculty of Agricultural Sciences, DAV University, Jalandhar, Punjab, India

A. Sharma, G. J. Ahammed (eds.), *Melatonin in Plants: Role in Plant Growth, Development, and Stress Response*, Plant Life and Environment Dynamics,
https://doi.org/10.1007/978-981-99-8051-2_8

along with the regulation of melatonin accumulation. An attempt has also been made to discuss the underlaying mechanisms and challenges for improving salt stress in different plant species.

Keywords Abiotic stress · Antioxidant enzymes · Salt stress tolerance · Melatonin · Transgenic expression · Seed priming · Bio-stimulants · Ion homeostasis

8.1 Introduction

Various abiotic stresses (drought, cold, flooding, UV, salt, heavy metal, water stress, etc.) can affect plant growth and development in a variety of ways. They can slow down the germination of seeds, slow development, encourage senescence, and even cause plant mortality. Salt stress is a major type of abiotic stressor that may severely affect the crop agricultural production, worldwide. Salinity affects agricultural production by affecting more than 400 million hectares of land, or over 6% of the world's total land area, globally (Munns and Tester 2008). Soil salinity drastically affects a number of morphological and physiological parameters of plants (Pitman and Läuchli 2006; Parida and Das 2005; Ahmad 2010). It not only affects plant growth and development but also severally affects various physiological, biochemical, and molecular processes that operate at the cellular to systemic level in the crop plants (Munns and Tester 2008; Cuartero et al. 2006; Nabati et al. 2011). The cellular-level processes include selective ion absorption and exclusion, ion compartmentalization within the central vacuole, production, and accumulation of organic solutes in the cytoplasm, and modifications to membrane composition (Abbasi et al. 2016; Rahnama et al. 2010; Ashraf and Wu 1994). The strategies at the whole-plant level include the modulation of photosynthetic pathway, activity of antioxidant enzymes, levels of plant hormones, ion absorption by roots, ion transport from roots to shoots, distribution of ions in shoots to various organs (such as old leaves and leaf sheaths), etc. (Shalata et al. 2001; Hasanuzzaman et al. 2018). Several studies have suggested that salt stress can lower K^+, Ca^{2+}, and Na^+ accumulation in different plant species, including canola (Ulfat et al. 2007), radish, cabbage (Jamil et al. 2007a, b), wheat (Raza et al. 2006), and sunflower (Akram et al. 2007), etc. Salinity lowers the nutrient availability and delivery to the plant's growing areas, which has an impact on the quality of both the plant's vegetative and reproductive parts. For instance, increased soil Na^+ concentrations may lead to a reduction of the Ca^{2+} activity in the external medium, which reduces the amount of Ca^{2+} available to *Celosia argentea* (Carter et al. 2005). Plants also respond to salt stress by producing hormones including ethylene, jasmonic acid, gibberellic acid, and abscisic acid, as well as other important signaling molecules like nitric oxide, hydrogen sulfide, and calcium, etc. Various researchers have tried to mitigate the harmful effects of salt stress using different bio-stimulants including some biomolecules, plant extracts or soil-inhabiting microorganisms. Out of which, a biomolecule "Melatonin" have been identified a potential candidate that can improve the plant adaptability toward the salt stress in a variety of ways and has a significant impact on how plants react to salt stress.

N-acetyl-5-methoxy-tryptamine, often referred to as melatonin, was initially discovered in 1958 and was named so as it counteracts the melanocyte-stimulating hormone's tendency to make the skin darker. Early melatonin research mostly focused on how it works in animals only (Ke et al. 2018). The impacts of several disorders, including coronary heart disease and Alzheimer's disease, as well as the activities of antioxidant enzymes, circadian rhythms, physical conditions, emotional state, and mood, are all regulated by melatonin in animals, either directly or indirectly (Lerner et al. 1958). In 1995, melatonin was found in the plants for the first time (Brainard et al. 2001; Mishima 2012). Since then, melatonin has been reported from a variety of plant species and from different plant organs, including roots, stems, leaves, fruits, and seeds, etc. Melatonin is a metabolite that has a wide array of roles in plants, including regulation of growth and development such as root organogenesis, flowering, and senescence. It is also found to be involved in the different responses of plants toward various stresses such as salinity, cold, drought, reactive oxygen species (ROS), and nutritional deficiencies (Mishima 2012; Rodriguez et al. 2004; Sharma and Zheng 2019; Sharma et al. 2020). Many researchers have recently concentrated on the function and regulation of melatonin-mediated responses of plants under salt stress. Melatonin-treated seeds for soybean (*Glycine max*) plants develop into strong seedlings with the larger leaves, greater plant height, increased plant biomass, more number of pods and gave higher seed yield than the untreated seeds. Furthermore, the melatonin treatment also boosted the tolerance of soybean plants toward both salt and drought stress (Imran et al. 2021). Bermuda grass (*Cynodon dactylon*) showed similar results, where the exogenous melatonin application enhanced plant tolerance toward salt, drought, and cold stress in comparison to the untreated plants (Shi et al. 2015a). Numerous studies have demonstrated that the melatonin-treated plants are better in terms of withstanding salt stress and exhibit reduced levels of ROS, electrolyte leakage, and cell damage; while having a higher plant height, biomass, and organic matter deposition than the untreated plants. Comparable outcomes were also observed in several other plants species under salt stress. Keeping this in mind, the present chapter is an attempt to discuss the use of melatonin in mitigating harmful effects of salt stress management via a different modes of melatonin treatment, and melatonin-mediated regulation of growth and development. Further an attempt has also been made to discuss its biosynthesis in plants, regulation of its biosynthesis, and the transgenic expression of the genes encoding the enzymes of its biosynthetic pathway genes for conferring a better tolerance to the salt stress have also been discussed.

8.2 Production of Endogenous Melatonin

Melatonin has been discovered to be present in a wide range of plants, including herbs, cultivated crops, fruit trees, etc. Melatonin is more likely to be synthesized in the mitochondria and chloroplasts of plants (Tan et al. 2013a; Chen et al. 2003;

Hernández-Ruiz and Arnao 2008a; Van Tassel et al. 2001; Byeon et al. 2012; Boccalandro et al. 2011). The concentration of melatonin documented by various groups in different plants or plant parts may be considered as the approximations only, as they are likely to be affected by seasonal and circadian rhythms in plants and therefore vary in both time and space (Hernández-Ruiz and Arnao 2008b). The distribution of endogenous melatonin in various organs and developmental stages of the plants are rather more comparable, despite the fact that the melatonin content varies substantially among different plant species.

Nature of stress, developmental phases as well as the circadian and seasonal cycles influence melatonin levels in plants (Kołodziejczyk et al. 2016a). The melatonin concentration often rises during maturation in tomatoes (*Lycopersicon esculentum* Mill.) and morning glory (Van Tassel et al. 2001). Different research revealed that environmental factors like light have an impact on the amount of endogenous melatonin in plants. Tomato plants raised in open fields had greater melatonin levels than those raised under the greenhouse conditions (Byeon et al. 2012). Furthermore, the melatonin level in senescing rice (*Oryza sativa*) leaves is higher in the presence of steady light than in its absence, suggesting that light signals also trigger the synthesis of melatonin during leaf senescence (Liang et al. 2015). However, in another study it was recorded that the light suppresses the production of melatonin in the grapevine (*Vitis vinifera*), as evidenced by the fact that its concentration declined sharply during the day and peaked at night (Boccalandro et al. 2011). Therefore, more detailed experimentation is required to find out the exact mechanism by which, light regulates the biosynthesis/accumulation of melatonin in different plant species.

8.3 Biosynthesis and Regulation of Melatonin in Plants

The biosynthesis of melatonin was first studied in animals (Yu et al. 1993; Olcese 2000; Reiter 1991). The aromatic amino acid tryptophan serves as its precursor. From tryptophan to melatonin, the entire biosynthetic pathway require four major enzymatic steps (Fig. 8.1). Tryptophan decarboxylase (TDC) catalyzes the conversion of tryptophan to tryptamine. Tryptamine 5-hydroxylase (T5H), the major pathway for serotonin biosynthesis in plants, then catalyzes the conversion of tryptamine to serotonin (Posmyk and Janas 2009). Another serotonin biosynthetic pathway, similar to the melatonin biosynthesis pathway in animals, is found in some plants like St. John's wort (*Hypericum perforatum*), where tryptophan 5-hydroxylase (TPH) converts tryptophan to 5-hydroxytryptophan, which is then catalyzed into serotonin by the action of aromatic-L-amino-acid decarboxylase (TDC/AADC) (Murch et al. 2000). Thereafter, serotonin is converted into *N*-acetyl-serotonin by the catalytic reaction of *N*-acetyltransferase (SNAT) or arylalkyl amine *N*-acetyltransferase (AANAT) followed by the conversion of *N*-acetyl-serotonin to melatonin by *N*-acetyl-serotonin methyl transferase (ASMT) or hydroxyl indole-*O*-methyl transferase (HIOMT). Additionally, SNAT can catalyze the

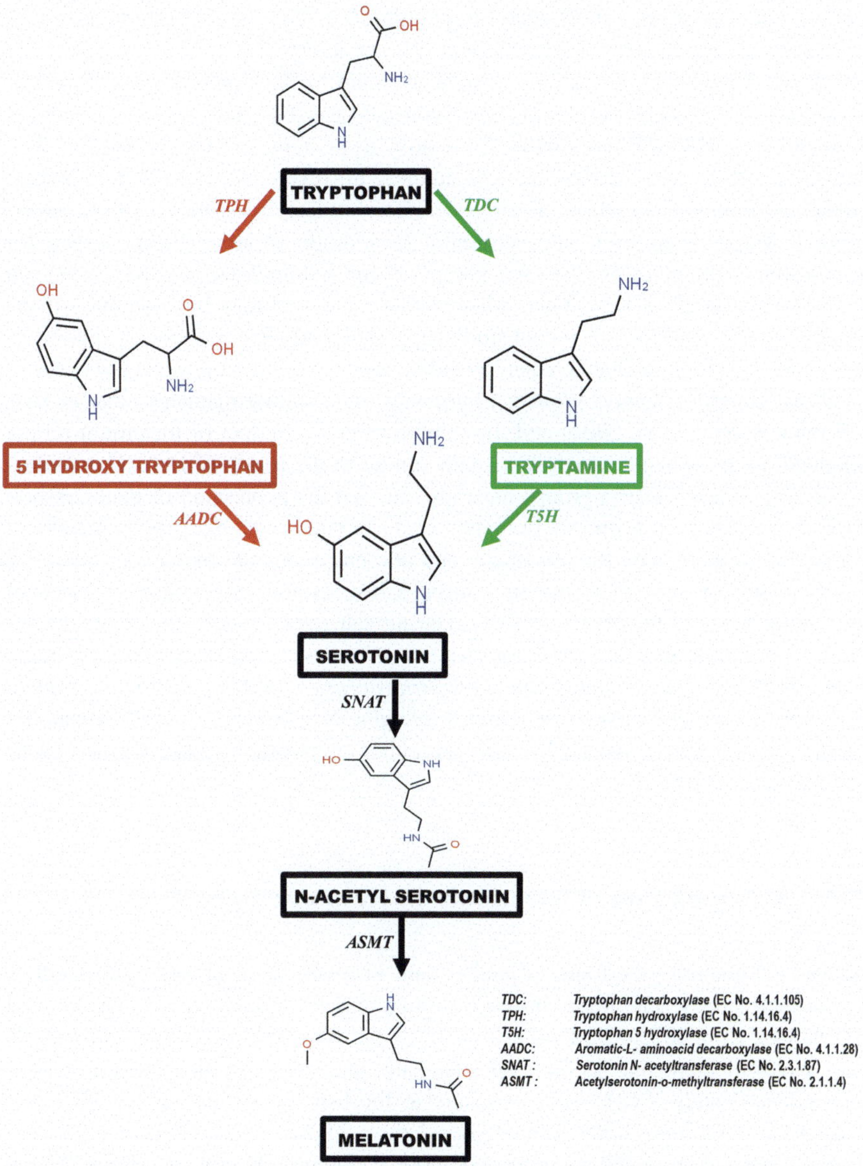

Fig. 8.1 Biosynthesis of melatonin: pathway in green color represents the reactions taking place in plants, pathway in red color represents the reactions taking place in animal, whereas pathway in black color represents the reactions taking place in both plant and animals

conversion of tryptamine to *N*-acetyl-tryptamine, but T5H cannot further convert *N*-acetyl-tryptamine into *N*-acetyl-serotonin (Zuo et al. 2014).

To our knowledge, no pathway has been found for the conversion of *N*-acetyl-tryptophan to *N*-acetyl-serotonin. Serotonin can also be converted to 5-methoxy tryptamine by HIOMT and, finally, 5-methoxy tryptamine is catalyzed by SNAT to melatonin (Tan et al. 2016). A few studies, however, also revealed a reverse melatonin pathway, in which *N*-acetyl-serotonin deacetylase catalyzes the conversion of *N*-acetyl-serotonin into serotonin. In addition to being the precursor for melatonin, tryptophan is also the precursor for indole-3-acetic acid (IAA), an auxin (Wang et al. 2012a). This indicates that melatonin may have similar effects in plants to those of IAA. A recent study suggested a possible, previously unknown, action for melatonin in plants, whereby indole amine promotes the vegetative growth in etiolated lupin (*Lupinus albus* L.) hypocotyls in a similar manner to that of IAA (Hernández-Ruiz et al. 2004) and the inhibitory growth effect of melatonin on the monocot roots assayed, which was again similar to the action of IAA (Hernández-Ruiz et al. 2005). Several groups have also studied the expression of genes associated to melatonin biosynthesis in response to salt stress with exogenous melatonin treatment, as well as on the enzymatic and non-enzymatic antioxidant systems. By the activation of antioxidant enzymes during salt stress, the exogenous application of melatonin reduced the levels of ROS. During salt stress, melatonin treatment has been recorded to upregulate the transcription of genes linked to melatonin production, including *N-acetylserotonin methyltransferase* (*ASMT1, ASMT2, ASMT3*), *tryptophan decarboxylase* (*TDC*), and *tryptamine 5-hydroxylase* (*T5H*). Melatonin controlled redox homeostasis by inducing either enzymatic or non-enzymatic antioxidant mechanisms.

8.4 Accumulation of Melatonin in Response to Stress

Melatonin functions as a special antioxidant in plants and interacts with ROS to lower their steady state levels in response to a variety of plant stresses (Arnao and Hernández-Ruiz 2013a, b). During salt stress, the amount of melatonin in the roots and leaves of grapevine seedlings dramatically increased, and the increase is dose dependent (Yandi 2018). After exposure to the osmotic stress, barley (*Hordeum vulgare* L.) and lupin both showed considerably higher melatonin levels. At high temperatures, rice seedlings produced more melatonin (Byeon and Back 2013). These findings suggest that stress-related situations trigger the manufacture of endogenous melatonin, suggesting that it participates in the plant's defense reaction to the diverse types of plant stresses. Melatonin synthesis and catabolic pathways-related gene expression and enzyme activities are directly correlated with melatonin accumulation in plants (Hardeland 2016). For instance, it was discovered that melatonin levels were closely correlated with a higher expression of genes involved in melatonin biosynthesis (like *TDC*, *T5H*, and *ASMT*, etc.) in rice grown under cadmium (Cd) stress (Cai et al. 2017). Moreover, enhanced SNAT and ASMT activity under high temperatures are linked to increased melatonin synthesis in rice (Byeon et al. 2015). Serotonin contributes to the stress response of rice under cold conditions, and melatonin concentration is typically, tightly correlated with the availability of its precursors. It was recorded that under cold and dry conditions, rice accumulates higher amounts of 2-hydroxymelatonin (Lee and Back 2016), which may play a role in the plant's ability to withstand these stresses. During Cd stress, direct binding of a transcription factor (HsfA1a) to the caffeic acid *O*-methyl transferase 1 (*COMT1*) gene promoter leads to an increase in melatonin levels in tomato plants. Under NaCl stress, the content of endogenous serotonin and melatonin was enhanced in both roots and cotyledons, demonstrating their active role in salt stress (Mukherjee et al. 2014). Subsequent investigation revealed that the enzyme responsible for the production of melatonin from *N*-acetyl-serotonin and *N*-acetyl-serotonin *O*-methyl transferase (ASMT) is modulated by NaCl stress (Mukherjee et al. 2014). However, the detailed regulation of the melatonin production under salt stress is, however, not fully understood yet.

8.5 Application of Various Mode of Exogenous Melatonin for Conferring Better Salt Stress Tolerance in Plants

Several researchers have suggested different possible roles of melatonin and the different modes of melatonin treatments. Melatonin treatment could stimulate root growth, encourage seed germination, and improve photosynthetic properties in a variety of crops (Martinez et al. 2018; Liang et al. 2019). Melatonin supplementation has been shown to increase a plant's ability to withstand salt stress and improve plant growth. The exogenous treatment of melatonin may supplement the endogenous melatonin level of melatonin in conferring salt stress tolerance in plants. Supplementation of plants with exogenous melatonin improved the ability watermelon and cucumber plants to withstand salt stress by promoting their growth and development under salinity stress (Li et al. 2017a, b; Wang et al. 2016a) Furthermore, seed priming has also been established as an effective method to increase yield and quality, uniformity, and germination rates in different plant species using a low dose of melatonin (Jabeen and Ahmad 2013; Mahmoudi et al. 2012; Ouhibi et al. 2014). Different workers have used different methods of exogenous melatonin application in plants such as seed soaking, foliar spray, supplementation of growth medium, seed priming, and along with water in a hydroponic system or under the field conditions, as listed in Table 8.1.

Table 8.1 Use of melatonin for salt stress amelioration in plants by its exogenous application in different modes

S. no.	Plant	Melatonin concentration	Mode of treatment	Effect	Reference
1	*Lactuca sativa*	50, 100, and 200 μm	Foliar spray	Decreased cell membrane damage, and improved chlorophyll content, RWC; decreased in the expression of *SOS1*, *SOS2*, *AKT1*, *NHX1*, and *HKT1* gene; increased DNA methylation	Taspinar et al. (2023)
2	*Stevia rebaudiana*	5, 100, and 500 μM	Seed priming	Increased in phenolic compound and all other phenolic acid	Simlat et al. (2023)
3	*Cynodon dactylon*	4, 20, and 100 μM/L	Irrigation with water	Improved salt, drought, and cold stress resistances by the alleviation of reactive oxygen species (ROS) burst and cell damage	Shi et al. (2015d)
4	*Ranunculus asiaticus*	50, 100, and 200 μM/L	Foliar application	Enhancing the plant growth traits such as chlorophyll and carotenoids content, relative water content, proline content, peroxidase enzyme activity (POD), and by the decreased electrolyte leakage rate, and Na^+ content, as well as delaying the emergence of flower buds	Eisa et al. (2023)
5	*Brassica oleracea*	50, 100 μM	Foliar spray	Improved morphological characteristics of broccoli, i.e., plant height, the number of leaves, head weight, head diameter, root length, shoot length, leaf relative water content, the number of florets, shoot fresh weight, root fresh, and dry weight	Sardar et al. (2023)

(continued)

Table 8.1 (continued)

S. no.	Plant	Melatonin concentration	Mode of treatment	Effect	Reference
6	*Abelmoschus esculentus*	100 μM MT	–	Increases the accumulation of proline and ratio of K^+/Na^+, photosynthetic pigment, relative water content, glutathione and ascorbic acid content and antioxidant enzymes content	Wang et al. (2023)
7	Snap Bean	20 μM and 100 μM	Irrigation with water	Reduction of oxidative damage and inhibition of polyamine catabolism	El-Beltagi et al. (2023)
8	*Triticum aestivum*	50, 100, 150, 200, 250, and 300 μM	Irrigation	Increased in antioxidant enzymes activity and gibberellin, jasmonic acid content	Wang et al. (2022)
9	*Cucumis sativus*	100 and 200 μM	Foliar spray	Increased protein content, essential nutrients (N, P, K^+, and Ca^{2+}) content, and catalase (CAT) activity and reduced levels of H_2O_2 and MDA	Brengi et al. (2022)
10	Rice	25, 50, 100, 200, 300, and 400 μM	Foliar spray	Improved antioxidant enzymes activity, reduced membrane lipid peroxidation, inhibited Na^+ uptake, alleviated cell injury in plant leaves, and increased N content and Si accumulation in the leaves	Wei et al. (2022b)
11	*Triticum aestivum*	70 μM	Foliar spray	Increased N, P, K^+, Fe, Zn, and Cu acquisition; enhanced ATP content and H^+-pump activity in roots; reduced superoxide radical content, electrolyte leakage, and lipoxygenase activity, as well as increased superoxide dismutase, catalase,	Talaat and Shawky (2022)

(continued)

Table 8.1 (continued)

S. no.	Plant	Melatonin concentration	Mode of treatment	Effect	Reference
				peroxidase, and polyphenol oxidase activities; K^+/Na^+, Ca^{2+}/Na^+, and Mg^{2+}/Na^+ ratios; relative water content; membrane stability index; and free amino acid accumulation	
12	*Carex leucochlora*	150 μmol/L	Foliar spray	Improved in the thickness of the leaf blade, vascular bundle area of the leaf main vein, vesicular cell area, thickness of the upper epidermis, and thickness of the lower epidermis	Ren et al. (2022)
13	*Phaseolus vulgaris* L.	100, 200 μM	Foliar spray	Improved weight of shoot, the root and the photosynthetic pigments, net photosynthesis rate, leaf stomatal conductance, transpiration rate, K^+, and Ca^{2+} content, and the ratio of K^+/Na^+; increased proline, sugar content and antioxidant enzyme activity, decreased malondialdehyde and H_2O_2 content	Azizi et al. (2022)
14	*Triticum aestivum*	70 μM	Foliar spray	Improved antioxidant enzymes activity, decreased toxic methylglyoxal content, up-regulation of the glyoxalase system	Talaat and Todorova (2022)
15	*Gossypium hirsutum*	10 μmol/L	Seed priming	Increased antioxidant enzymes activity, soluble sugar, soluble protein, and K^+ ion content, decreased accumulation of	Duan et al. (2022)

(continued)

Table 8.1 (continued)

S. no.	Plant	Melatonin concentration	Mode of treatment	Effect	Reference
				H_2O_2 and MDA content	
16	*Poncirus trifoliata*	50, 100, 150 μmol/L	Irrigation	Improved root zeatin riboside, gibberellin and brassinolide content, plant growth, root auxin levels, leaf photosynthesis	Hu et al. (2022)
17	*Musa acuminata*	100 μM	Irrigation	Enhanced proline, chlorophyll, free protein, RWC content and reduced MDA content	Wei et al. (2022a)
18	*Solanum lycopersicum*	50, 100 μM	Foliar spray	Improved photosynthesis, proline content, cell viability, stomatal behavior, plant growth, and antioxidant enzyme activity, and decrease reactive oxygen species, MDA content	Khan et al. (2022)
19	*Triticum aestivum*	200 μM	Irrigation	Improved photosynthesis and fresh weight	Ahmad et al. (2022)
20	*Madicago sativa*	0.1, 10, 15 μM	Irrigation	Enhanced the expression of antiapoptotic genes *BI-1, UCP1-UCP2* decreased the expression of proapoptotic gene *γVPE*, and decreased H_2O_2, O_2^- content	Jalili et al. (2022)
21	*Zea mays*	0.8 mM	Seed priming	Improved antioxidant enzymes activity, relative water content, proline and total phenolic content, decreased electrolyte leakage, Na^+ ion content, and lipid peroxidation product	Jiang et al. (2016)
22	*Vicia faba*	100 and 150 mM	Seed priming	Enhanced carbohydrate content, photosynthetic efficiency, indole acetic acid content, phenolic	Dawood et al. (2015)

(continued)

Table 8.1 (continued)

S. no.	Plant	Melatonin concentration	Mode of treatment	Effect	Reference
				content, and K^+ and Ca^{2+} ion content	
23	*Glycine max*	0, 50, 100 μM	Seed priming	Increased fatty acid content, number of pods and seeds	Wei et al. (2015)
24	*Cucumis sativus*	1 μM	Seed priming	Promoted catabolism of ABA and biosynthesis of gibberellic acid and recharged the antioxidant system	Zhang et al. (2014a, b)
25	*Malus hupehensis*	0.1 μM	Given hydroponically in media	Increased the expression of *MdNHX1* and *MdAKT1* genes, and antioxidant enzymes	Li et al. (2012a)

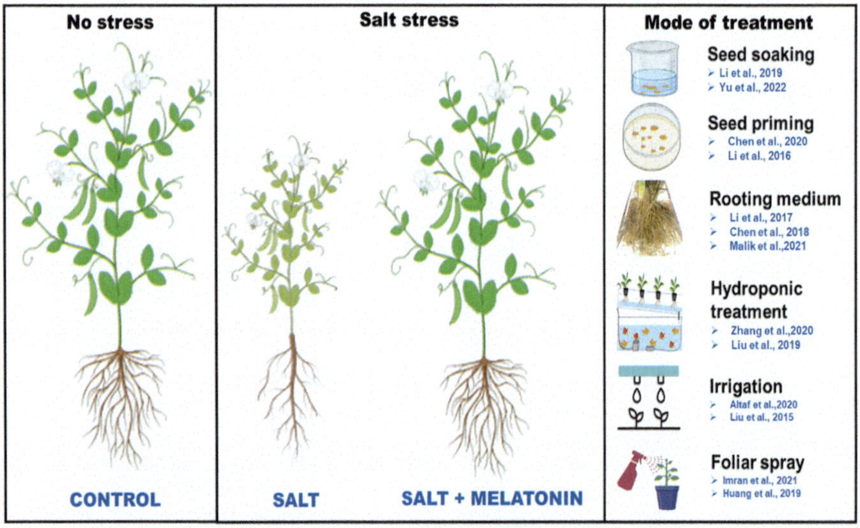

Fig. 8.2 Various modes of exogenous application of melatonin for ameliorating harmful effects of salt stress in plants and conferring better salt stress tolerance

8.5.1 Seed Soaking

Seeds of some plant species take a long time to germinate, soaking is a process that accelerates the germination process and breaks the dormancy of seeds (Fig. 8.2). Various concentrations of melatonin have been tried for seed soaking treatments in different plants like in cotton (Duan et al. 2022) and rice (Li et al. 2017a, b; Huangfu et al. 2021).

8.5.2 Seed Priming

Seed priming is the process of controlled hydration of seeds to a level that permits pre-germinative metabolic activity to proceed but prevents the actual emergence of the radicle. Seed priming with melatonin has been used in some plants such as *Stevia rebaudiana* (Simlat et al. 2023) and *Zea maize* (Jiang et al. 2016) under salt stress.

8.5.3 Growth Medium

Any growing medium (with several nutritional constituents), used to propagate plants (from seeds, cuttings or other propagules) is known as a growth medium. Different workers have tried melatonin treatment in the growth medium to amlerate the harmful effects of salt stress. Melatonin enhances antioxidant activity, ion homeostasis, and root growth in canola seedlings (Javeed et al. 2021) and ameliorates salt stress in melon and tomato (Castañares and Bouzo 2019; Ali et al. 2021).

8.5.4 Hydroponic Treatment

In this method plants are grown in nutrient-rich solution without soil. In rice, melatonin treatments effectively delayed leaf senescence, lowered the transcripts of senescence-associated genes, decreased chlorophyll degradation, and improved salt stress resistance (Liang et al. 2015), when given hydroponically.

8.5.5 Irrigation with Water

It refers to the application of melatonin through irrigation water. In this method, melatonin is carried into the soil in a solution that is absorbed by the roots of the plant. Melatonin irrigation with water has been used in some plants such as *Cynodon dactylon* (Li et al. 2015), and Snap Bean (El-Beltagi et al. 2023) and was found effective against salt stress.

8.5.6 Spraying Method

It refers to the spraying of melatonin solution on the foliage of growing plants. In the spraying method, the melatonin is mixed in methanol, and then melatonin mixed with methanol is dissolved in water because melatonin does not mix directly in water. Melatonin mixture sprayed on leaves of plants because foliar application of melatonin is effective and the concentration of the spray solution has to be controlled, otherwise serious damage may result due to scorching of the leaves. Spraying method has been used by many researchers in different plants such as *Lactuca sativa* (Taspinar et al. 2023), *Brassica oleracea* (Sardar et al. 2023), *Oryza sativa* (Wei et al. 2022a), etc.

8.6 Various Functions of Exogenous Melatonin Under Salt Stress

Melatonin and IAA are structurally related since they are both indole amines and have the same biosynthetic precursor so they affect similarly on plant development and growth (Fig. 8.3). Exogenous melatonin treatment reduced the release of reactive oxygen species and maintained photosynthetic activity under salt stress by activating antioxidant enzymes (Chen et al. 2018).

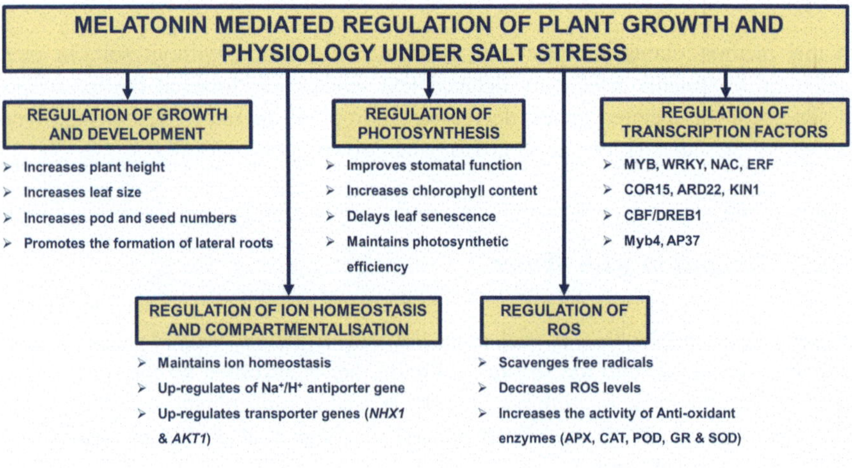

Fig. 8.3 Different regulatory effects of exogenous melatonin in improving salt stress tolerance in plants

8.6.1 Regulation of Growth and Development

Melatonin plays a similar role to auxin in many crops, including canary grass (*Phalaris canariensis*), barley, wheat (*Triticum* sp.), and oat (*Avena sativa*) (Hernández-Ruiz et al. 2005), as well as dicots such as Arabidopsis thaliana and lupin (*Lupinus micranthus*). Melatonin is structurally similar to IAA (Hernández-Ruiz et al. 2005; Chen et al. 2017). Exogenous melatonin significantly affects plant growth and development when melatonin levels are low in plants (Kolár et al. 1997). Melatonin that has been exogenously given is permeable across the plasma membrane and raises the level of endogenous melatonin, which has been demonstrated to encourage soybean plant growth and seed output (Wei et al. 2015). Exogenous melatonin also aids early flower growth, which is influenced by photoperiod (Kolár et al. 2003). Exogenous melatonin treatment may also encourage the development of lateral roots in cucumber (Zhang et al. 2014a, b). It can be applied via various mode of treatment to plants (Fig. 8.3) at an appropriate concentration, and the optium exogenous melatonin concentration may be worked on plant-to-plant bases as different workers have reported different concentrations in different plant species (Table 8.2). A moderate melatonin dosage in *Arabidopsis* seedlings encourages plant development, while a low concentration has no significant impact. However, a high concentration may inhibits plant growth too (Bajwa et al. 2014). Therefore, the researchers need to be very careful in selecting appropriate melatonin concentration for salt stress amelioration.

8.6.2 Regulation of Photosynthesis Under Salt Stress

In photosynthesis and photoprotection, melatonin plays a crucial role (Wang et al. 2012b). By lowering the chlorophyll content, the actual photochemical efficiency of PSII, and photochemical quenching (qP), salt stress restricts the absorption of light energy and electron transport in photosystem II (PSII), which has a negative impact on the bioenergetic process of photosynthesis (Hao et al. 2017; Lin et al. 2018; Takahashi and Murata 2008). Treatment with melatonin decreased the impact of salt stress on the biomass and photosynthetic processes. According to research on the macroalga *Ulva* sp. (Tal et al. 2011) and the freshwater *Chara australis*, melatonin was recorded to exert the protective effects on chlorophyll that boost the effectiveness of the reaction centers of photosystem II (Lazár et al. 2013). Similar results were recently also found in salt-stressed bermuda grass, citrus, and sunflower supported the melatonin's protective effects on photosynthetic pigments (Shi et al. 2015a). Under salt stress, plants close their stomata to prevent water loss, which lowers stomatal conductance (GS), which in turn lowers photosynthesis (Brugnoli and Lauteri 1991; Meloni et al. 2003); yet, melatonin can enhance stomatal function and allow plants to reopen their stomata in response to salt stress when used in an appropriate quantity (Ye et al. 2016). Melatonin administration reduces the potential

Table 8.2 Transgenic expression of some key genes involved in melatonin biosynthesis to improve salt stress tolerance in plants.

S. no.	Genes	Source organism	Expressed in plant	Effect	Reference
1	*VvSNAT1*	Grape	*Arabidopsis*	Promote the growth of plants and reduce oxidative damage	Wu et al. (2021)
2	*SlCOMT1* gene	Tomato	*Solanum lycopersicum*	Enhance salt tolerance in tomato	Liu et al. (2019)
3	*HIOMT* gene *MdASMT* *MdNCED3*	Animal	*Malus domestica*	Maintaining high photosynthetic capacity, reducing ROS accumulation, maintaining ion homeostasis	Tan et al. (2021)
4	*oAANAT* and *oHIOMT*	Sheep	*Panicum virgatum*	Promoting flowering, growth, and salt tolerance	Huang et al. (2017)
5	*MnT5H2*	Mulberry	*Nicotiana tabacum*	Improve salt tolerance	Zhu et al. (2022)
6	*CsSNAT* gene	Tea plant	*Cucumis sativus*	Improve growth and salt tolerance	Qi et al. (2020)
7	*oAANAT* and *oHIOMT*	Sheep	*Solanum lycopersicum* variety micro-tom	Enhance drought tolerance	Wang et al. (2014)
8	*MzASMT 1*	Apple	*Nicotiana tabacum*	Improving salt stress tolerance	Zhuang et al. (2020)
9	*SlCOMT1*	Tomato	*Solanum lycopersicum*	Improving salt stress tolerance	Sun et al. (2020)
10	*MzASMT1* gene	Apple	*Arabidopsis*	Enhance drought tolerance	Zuo et al. (2014)
11	*AANAT* or *HIOMT*	Animal	*Malus domestica*	Enhance UV- tolerance	Liu et al. (2021)

of the net photosynthetic rate (Pn) to rapidly decline during salt stress. By increasing chlorophyll production and reducing the rate of chlorophyll degradation during salt stress, melatonin administration increased the maximum photochemical. Melatonin is essential for protecting PSII and preventing the loss of chlorophyll content caused by salt stress. Melatonin improved the expression of photosynthesis-related genes in bermuda grass (*Cynodon dactylon*) during salt stress and improved glucose metabolism, fatty acid metabolism, and ascorbic acid production (Shi et al. 2015b). Under the influence of melatonin, the gene expression of the PSI-related proteins PsaK and PsaG and their components PsbO and PsbP, as well as the PSII photochemical reaction center protein OEC (oxygen-evolving enhancer proteins) was up-regulated. Moreover, under salt stress, melatonin up-regulated the transcription of genes involved in photosynthesis and shielded the photosynthetic system (Shi et al. 2015a).

8.6.3 Regulation of Transcription Factors Under Salt Stress

Melatonin regulates the activity of some key transcription factors, which may improve the salt tolerance in plants. Zinc finger protein 6 (ZAT6), heat shock factors (HSFA1s), and C-repeat-binding factor/drought response element binding 1 factors (CBF/DREB1s) are the three primary melatonin-associated transcription factors in plants. CBF/DREB1, which is closely associated with high levels of melatonin, simultaneously up-regulates a number of stress response genes, such as *cold-inducible 1* (*KIN1*), *cold-related 15A* (*COR15A*), and responsive to *dehydration 22* (*RD22*), increasing plant tolerance toward salt, drought, and freezing stresses (Shi et al. 2015c). 2-hydroxymelatonin in rice upregulates Myb4 and AP37 transcription factor during abiotic stress (Kang et al. 2010). In apple, the transcript of chlorophyll degradation gene *pheophorbide an oxygenase* (*PAO*), *senescence-associated gene 12* (*SAG12*), and *auxin resistant 3* (*AXR3*)/*indole-3-acetic acid inducible 17* (*IAA170*) is downregulated by melatonin. The 77 differentially expressed genes, including some significant transcription factors, were upregulated by melatonin (e.g., *MYB, WRKY, NAC*, and *ERF*) treatment. The ability of cucumbers to tolerate salt was also found to be closely correlated with the upregulation of these transcription factors (Wei et al. 2018).

8.6.4 Regulation of Ion Homeostasis and Their Compartmentalization

An excessive amount of major intracellular salt ions (Na^+, K^+, Ca^{2+}, and H^+) disrupts the mechanism of plant and affects the growth and development of the plant (Amtmann and Sanders 1998; Zhu 2003). Exogenous melatonin maintains ion homeostasis by upregulating various genes. Under saline conditions plant absorbs high amount of Na^+ ion which affects cytosolic enzymes. Thus, maintaining high K^+ and low Na^+ cytosolic levels by K^+ and Na^+ concentration modulation has a big impact on salt-stressed plants (Zhu 2003; Fukuda et al. 2011). There are three main processes that inhibit Na^+ accumulation in the cytoplasm: limiting Na^+ input, active Na^+ outflow, and compartmentalizing Na^+ into the vacuole (Padan et al. 2001). A vacuolar Na^+/H^+ exchanger is encoded by the *NHX1* gene, whose *Arabidopsis* counterpart, *AtNHX1*, was increased by salt stress, resulting in an excess transport of Na^+ into vacuolar (Shi and Zhu 2002). Plasma membrane Na^+/H^+ antiporter is the transmembrane protein that *Salt Overly Sensitive1* (*SOS1*) gene encodes. The Na^+ is transported out of the cells by SOS signaling (Zhao et al. 2010; Padan et al. 2001). Melatonin significantly increased K^+ in shoots of maize seedlings (Jiang et al. 2016) as well as in melatonin pretreated *Malus hupehensis* seedlings (Li et al. 2012a). In melatonin-treated rapeseed seedlings, *NHX1* and *SOS2* expression was found to be higher (Zhao et al. 2010). However, there is little evidence that Ca^{2+} signaling is involved in melatonin-triggered salinity tolerance. However, the Ca^{2+} signaling is known to play a crucial role under various types of both biotic and abiotic stress responses in plants, including the salt stress.

8.6.5 *Regulation of Enzymatic and Non-Enzymatic Antioxidants During Salt Stress*

Aa result of salt stress, more Reactive Oxygen Species (ROS) are accumulated in the plants, which damages the cells and triggers other defense reactions (Guo et al. 2018a; Blokhina et al. 2003; Li et al. 2012a, b; Duan et al. 2018; Sairam and Srivastava 2002). Melatonin and traditional antioxidants like vitamin C, vitamin E, and glutathione are examples of non-enzymatic antioxidants that are part of plants' ROS scavenging mechanisms. Enzymatic antioxidants are the enzymes such as catalase and superoxide dismutase, etc., which scavenge ROS as a part of their enzyme reactions. Compared to the other non-enzymatic antioxidants, melatonin demonstrates a stronger antioxidant potential (Tan et al. 2013a, b). Melatonin, which is both produced naturally and may be given exogenously, can successfully reduce the oxidative damage caused by ROS. According to estimates, melatonin neutralizes ROS through a cascade process (Tan et al. 2013a, b). Research revealed that plants pretreated with melatonin contained comparatively little H_2O_2. Hence, melatonin's antioxidant properties have drawn attention, and it is generally accepted that its primary purpose is to directly scavenge ROS. Furthermore, enzyme-based antioxidants including superoxide dismutase (SOD), catalase (CAT), ascorbate oxidase (APX), glutathione peroxidase (GPX), and glutathione reductase offer plants a very effective and targeted ROS scavenging system (GR). When plants are exposed to salt stress, the activities of these antioxidant enzymes tend to increase, and their levels are correlated with the plant's salt tolerance (Mittler 2002; Guo et al. 2018b). Researchers revealed that compared to untreated seedlings, melatonin-pretreated seedlings had increased antioxidant enzyme activity (Tan et al. 2002). The common view is that the melatonin may enhance the activity of antioxidant enzymes. Moreover, melatonin can boost the mitochondrial electron transport chain's effectiveness, reducing the free radical production and easing the electron leakage, which protects antioxidant enzymes from the oxidative damage (Parida and Das 2005).

8.6.6 *Melatonin Metabolism-Related Genes and Their Expression in Plants Under Salt Stress*

Tryptophan, which is converted to tryptamine by tryptophan decarboxylase, is the first step in the melatonin biosynthesis pathway in a variety of plant species. Tryptamine 5-hydroxylase then converts tryptamine to serotonin (T5H). Serotonin is then converted to *N*-acetylserotonin by the enzyme serotonin *N*-acetyltransferase

(SNAT), while melatonin is produced from *N*-acetylserotonin by the enzymes *N*-acetylserotonin methyl transferase (ASMT) or caffeic acid *O*-methyltransferase (COMT). Given that SNAT's substrate affinity is for 5-methoxy tryptamine and ASMT/substrate COMT's affinity is for serotonin, it is possible that serotonin was first methylated to 5-methoxy tryptamine by ASMT/COMT, before being converted to melatonin by SNAT. The different phases of the melatonin production pathways may take place in several subcellular sites. When *VvSNAT1* from grapes was overexpresses in transgenic *Arabidopsis thaliana*, it promoted the growth of plants and reduced oxidative damage (Table 8.2) by reducing MDA and H_2O_2 and improving salt tolerance (Wu et al. 2021). In cucumber, over-expression of *CsSNAT* positively regulated salt stress tolerance, lowered the membrane damage, and decreased the accumulation of ROS (Qi et al. 2020).

8.7 Cross-Talk of Melatonin with Other Phytohormones

Melatonin also plays an important role in the regulation of various other phytohormone (like, indole-3-acetic acid (IAA), abscisic acid (ABA), gibberellic acids (GA), cytokinins (CK), and ethylene, etc.), networks (Arnao and Hernández-Ruiz 2018). It is crucial for regulating, directing, and balancing more traditional phytohormone signaling cascades, with the specific action and response differing according to the function (Erland et al. 2015; Arnao and Hernández-Ruiz 2018, 2021). Melatonin is now known to be found in plants also, but its function, signaling, or interactions with other phytohormone is still under investigation. Tryptophan is the substrate for both melatonin as well as an auxin, i.e., IAA (Wang et al. 2016b). *Brassica juncea* (Arnao and Hernández-Ruiz 2018; Chen et al. 2009) was shown to slightly increase its endogenous IAA concentration, after it was exposed to the exogenous melatonin, supporting the theory that melatonin regulates the development of lateral roots also, in Arabidopsis (Pelagio-Flores et al. 2012). Additionally, the treatment of IAA in low quantities raises levels of endogenous melatonin too. At the same time, high melatonin concentrations suppress *PIN1,3,7* expression and lower IAA levels in Arabidopsis roots, indicating that melatonin, in whole or in part, may control the root growth in Arabidopsis through auxin production and polar auxin transport (Wang et al. 2016b).

Melatonin also raised endogenous ABA content in *Elymus nutans* after salt stress; however, fluridone dramatically reduced this effect (Fu et al. 2017). The concentration of endogenous melatonin was although remain unaffected by pretreatments with ABA and fluridone. By controlling ABA production and its catabolism, exogenous melatonin can assist to decrease the effects of salt stress. The transcripts levels of ABA synthesis-related genes *CsNCED1* and *CsNCED2* were decreased during salt stress in melatonin-pretreated seeds, but genes involved in ABA catabolism had their transcript levels significantly increased, as a result, the amount of ABA was reduced. On the other hand, melatonin significantly increased the expression of the GA synthesis genes *GA20ox* and *GA3ox*, which were associated with the elevated GA content (Zhang et al. 2014a, b). In general, more detailed studies on the biosynthesis and catabolism of different PGRs in order to understand its interplay with the other PGRs in a better way in under salt stress.

8.8 Effect of Melatonin on Plant Proteome During Salt Stress

Proteomic analysis can be used to identify proteins that are regulated by exogenous melatonin during certain biological reactions or in specific processes. Melatonin and 309 proteins, the majority of which have hydrolase activity, have been shown to govern the senescence of *Malus hupehensis* leaves (Wang et al. 2014). Heat shock proteins (HSPs) are produced by plants as a crucial adaptation mechanism in response to a variety of stresses (Timperio et al. 2008), including the salt stress. By restoring stress-damaged proteins, HSPs function as molecular chaperones to assist in defending cells against stress (Vierling 1991; Sun et al. 2002). NaCl stress has been linked to HSP70. In cucumber melatonin regulated 11 HSPs to protect seed germination under salt stress (Zhang et al. 2017). The accumulation of ROS such $O_2{\bullet}-$, H_2O_2, and •OH is brought on by salt stress (Zhang et al. 2013, 2014a, b; Hasegawa et al. 2000). By scavenging ROS, melatonin has been shown to reduce the inhibitory effect of excessive salt on the germination of cucumber seeds (Zhang et al. 2014a, b). The expression of enzymes CsCu-ZnSOD, CsFe-ZnSOD, CsPOD, and CsCAT were considerably increased after receiving exogenous melatonin therapy, and oxidative damage was reduced (Zhang et al. 2014a, b). During salt stress, the exogenous melatonin treatment increased the GPXs, SODs, and PODs from 2.1- to 10.2-fold (Zhang et al. 2017). Glycolysis, citric acid cycle, glyoxylate cycle, stress response, and cell elongation were all regulated by exogenous melatonin (Firenzuo et al. 1968), and the responces were thought to be mediated by different enzyme activities. During seed germination, glycolysis and the TCA cycle enzymes play a crucial role in supplying energy for cell activity and the carbon skeleton needed for biomolecule production (Xu et al. 2016a, b). By controlling energy production, through the enzymes of carbohydrate metabolism, melatonin primarily alleviates the inhibitory effect of excessive salt on the germination of cucumber seeds (Zhang et al. 2017).

8.9 Effect of Melatonin on Transcriptomics During Salinity

Transcriptomics gives gene transcription maps and information on how they are regulated, elucidates the molecular basis of diverse biological factors, and clarifies the way that genetic and environmental factors interact (Cho et al. 2001; Zhang et al. 2019). In okra, the exogenous application 50-μM melatonin enhanced more than

10 GO terms and 34 KEGG pathways, and nitrogen metabolism, sulfur metabolism, and alanine, aspartate, and glutamate metabolism were also enhanced. Some transcription factors including *MYB, WRKY, NAC*, etc. were also identified as differentially expressed genes (Zhan et al. 2021). Melatonin affects plant metabolic equilibrium via nitrogen metabolism, hormone metabolism, and the tricarboxylic acid cycle as recorded in a transcriptome analysis by Wang et al. (2019).

8.10 Conclusion and Future Prospects

The understanding of melatonin's functions in plants has improved over time. This research has added to our understanding of melatonin's levels and distribution, as well as its metabolism and role in plants. The signaling mechanism of melatonin during salt stress is still unknown. Melatonin enhances stress-related transcription factors, antioxidant enzyme activity, photosynthetic efficiency, and metabolite concentration. This review summarizes the mechanisms by which melatonin alleviates salt-induced damage and discusses the melatonin-mediated regulation of growth and development and salt stress signaling in plants. It is remarkable that additional melatonin precursors and metabolite chemicals may improve a plant's capacity to reduce salt stress and improve crop yield under salt stress. Overall, melatonin has the potential to be a bio-stimulant that can help plants improve their resistance to salt stress. It can also be used more successfully in various plants to manage salt stress. In conclusion, melatonin's ability to improve plant adaptability under salt stress can make it a potential molecule to mitigate adverse effects of salt stress in plants. However, a detailed studies using various omics approaches and careful optimization of its doses and mode of application on plant-to-plant based will provide more useful information to access in large scale application under field conditions.

Acknowledgments The authors are thankful to the DAV University administration for providing all necessary infrastructural facilities and seed money for research. Dr. Rahul Kumar and Dr. Ashutosh Sharma acknowledge the financial grant received from the Council of Industrial Scientific Research (CSIR-HRDG), New Delhi, India, through an extramural research grant with project number 38(1477)/19/EMR-II, to work on salt stress.

Conflict of Interest The authors declare that they do not have any conflict of competing interests.

References

Abbasi H, Jamil M, Haq A, Ali S, Ahmad R, Malik Z, Parveen (2016) Salt stress manifestation on plants, mechanism of salt tolerance and potassium role in alleviating it: a review. Zemdirbyste 103(2):229–238. https://doi.org/10.13080/z-a.2016.103.030

Ahmad P (2010) Growth and antioxidant responses in mustard (*Brassica juncea* L.) plants subjected to combined effect of gibberellic acid and salinity. Arch Acker Pflanzenbau Bodenkd 56(5):575–588. https://doi.org/10.1080/03650340903164231

Ahmad I, Munsif F, Mihoub A, Jamal A, Saeed MF, Babar S, Fawad M, Zia A (2022) Beneficial effect of melatonin on growth and chlorophyll content in wheat (*Triticum aestivum* L.) grown under salt stress conditions. Gesunde Pflanzen 74(4):997–1009. https://doi.org/10.1007/s10343-022-00684-5

Akram MS, Athar HR, Ashraf M (2007) Improving growth and yield of sunflower (*Helianthus annuus* L.) by foliar application of potassium hydroxide (KOH) under salt stress. Pak J Bot 39(3):769–776

Ali M, Kamran M, Abbasi GH, Saleem MH, Ahmad S, Parveen A, Malik Z, Afzal S, Ahmar S, Dawar KM, Ali S, Alamri S, Siddiqui MH, Akbar R, Fahad S (2021) Melatonin-induced salinity tolerance by ameliorating osmotic and oxidative stress in the seedlings of two tomato (*Solanum lycopersicum* L.) cultivars. J Plant Growth Regul 40(5):2236–2248. https://doi.org/10.1007/s00344-020-10273-3

Amtmann A, Sanders D (1998) Mechanisms of Na^+ uptake by plant cells. In: Advances in botanical research. Elsevier, Amsterdam, pp 75–112

Arnao MB, Hernández-Ruiz J (2013a) Growth conditions determine different melatonin levels in *Lupinus albus* L. J Pineal Res 55(2):149–155. https://doi.org/10.1111/jpi.12055

Arnao MB, Hernández-Ruiz J (2013b) Growth conditions influence the melatonin content of tomato plants. Food Chem 138(2–3):1212–1214. https://doi.org/10.1016/j.foodchem.2012.10.077

Arnao MB, Hernández-Ruiz J (2018) Melatonin and its relationship to plant hormones. Ann Bot 121(2):195–207. https://doi.org/10.1093/aob/mcx114

Arnao MB, Hernández-Ruiz J (2021) Melatonin as a regulatory hub of plant hormone levels and action in stress situations. Plant Biol (Stuttg) 23(Suppl 1(S1)):7–19. https://doi.org/10.1111/plb.13202

Ashraf MY, Wu L (1994) Breeding for salinity tolerance in plants. Critical Reviews in Plant Sciences 13(1):17–42.

Azizi F, Amiri H, Ismaili A (2022) Melatonin improves salinity stress tolerance of Phaseolus vulgaris L. cv. Pakby changing antioxidant enzymes and photosynthetic parameters. Acta Physiol Plant 44(4):40

Bajwa VS, Shukla MR, Sherif SM, Murch SJ, Saxena PK (2014) Role of melatonin in alleviating cold stress in *Arabidopsis thaliana*. J Pineal Res 56(3):238–245. https://doi.org/10.1111/jpi.12115

Blokhina O, Virolainen E, Fagerstedt KV (2003) Antioxidants, oxidative damage and oxygen deprivation stress: a review. Ann Bot 91(2):179–194

Boccalandro HE, González CV, Wunderlin DA, Silva MF (2011) Melatonin levels, determined by LC-ESI-MS/MS, fluctuate during the day/night cycle in *Vitis vinifera* cv Malbec: evidence of its antioxidant role in fruits: melatonin day/night cycle fluctuation in *Vitis vinifera*. J Pineal Res 51(2):226–232. https://doi.org/10.1111/j.1600-079X.2011.00884.x

Brainard GC, Hanifin JP, Greeson JM, Byrne B, Glickman G, Gerner E, Rollag MD (2001) Action spectrum for melatonin regulation in humans: evidence for a novel circadian photoreceptor. J Neurosci 21(16):6405–6412. https://doi.org/10.1523/jneurosci.21-16-06405.2001

Brengi SH, Khedr AAEM, Abouelsaad IA (2022) Effect of melatonin or cobalt on growth, yield and physiological responses of cucumber (*Cucumis sativus* L.) plants under salt stress. J Saudi Soc Agric Sci 21(1):51–60. https://doi.org/10.1016/j.jssas.2021.06.012

Brugnoli E, Lauteri M (1991) Effects of salinity on stomatal conductance, photosynthetic capacity, and carbon isotope discrimination of salt tolerant (*Gossypium hirsutum* L.) and salt-sensitive (*Phaseolus vulgaris* L.) C3 non-halophytes. Plant Physiol 95:628–635

Byeon Y, Back K (2013) Melatonin synthesis in rice seedlings in vivo is enhanced at high temperatures and under dark conditions due to increased serotonin N-acetyltransferase and N-acetyl serotonin methyl transferase activities. J Pineal Res 56:189–195

Byeon SE, Yi Y-S, Oh J, Yoo BC, Hong S, Cho JY (2012) The role of Src kinase in macrophage-mediated inflammatory responses. Mediators Inflamm 2012:512926. https://doi.org/10.1155/2012/512926

Byeon Y, Lee HY, Hwang OJ, Lee H-J, Lee K, Back K (2015) Coordinated regulation of melatonin synthesis and degradation genes in rice leaves in response to cadmium treatment. J Pineal Res 58(4):470–478. https://doi.org/10.1111/jpi.12232

Cai SY, Zhang Y, Xu YP, Qi ZY, Li MQ, Ahammed GJ, Xia XJ, Shi K, Zhou YH, Reiter RJ, Yu JQ (2017) HsfA1a up regulates melatonin biosynthesis to confer cadmium tolerance in tomato plants. J Pineal Res 62(2):e12387

Carter CT, Grieve CM, Poss JA, Suarez DL (2005) Production and ion uptake of *Celosia argentea* irrigated with saline wastewaters. Sci Hortic 106(3):381–394. https://doi.org/10.1016/j.scienta.2005.04.007

Castañares JL, Bouzo CA (2019) Effect of exogenous melatonin on seed germination and seedling growth in melon (*Cucumis melo* L.) under salt stress. Hortic Plant J 5(2):79–87

Chen G, Huo Y, Tan DX, Liang Z, Zhang W, Zhang Y (2003) Melatonin in Chinese medicinal herbs. Life Sci 73(1):19–26. https://doi.org/10.1016/s0024-3205(03)00252-2

Chen Q, Qi WB, Reiter RJ, Wei W, Wang BM (2009) Exogenously applied melatonin stimulates root growth and raises endogenous indole acetic acid in roots of etiolated seedlings of *Brassica juncea*. J Plant Physiol 166:324–328

Chen L, Fan J, Hu Z, Huang X, Amombo E, Liu A, Bi A, Chen K, Xie Y, Fu J (2017) Melatonin is involved in regulation of Bermuda grass growth and development and response to low K^+ stress. Front Plant Sci 8:2038

Chen Y-E, Mao J-J, Sun L-Q, Huang B, Ding C-B, Gu Y, Liao J-Q, Hu C, Zhang Z-W, Yuan S, Yuan M (2018) Exogenous melatonin enhances salt stress tolerance in maize seedlings by improving antioxidant and photosynthetic capacity. Physiol Plant 164(3):349–363. https://doi.org/10.1111/ppl.12737

Cho RJ, Huang M, Campbell MJ, Dong H, Steinmetz L, Sapinoso L, Hampton G, Elledge SJ, Davis RW, Lockhart DJ (2001) Transcriptional regulation and function during the human cell cycle. Nat Genet 27(1):48–54. https://doi.org/10.1038/83751

Cuartero J, Bolarín MC, Asíns MJ, Moreno V (2006) Increasing salt tolerance in the tomato. J Exp Bot 57(5):1045–1058. https://doi.org/10.1093/jxb/erj102

Dawood MG, El-Awadi ME, Duan W, Lu B, Liu L, Meng Y, Ma X, Li J (2015) Effects of exogenous melatonin on root physiology, transcriptome and metabolome of cotton seedlings under salt stress. Int J Mol Sci 20(2):9456

Domingos P, Prado AM, Wong A, Gehring C, Feijo JA (2015) Nitric oxide: a multitasked signaling gas in plants. Mol Plant 8(4):506–520. https://doi.org/10.1016/j.molp.2014.12.010

Duan H, Ma Y, Liu R, Li Q, Yang Y, Song J (2018) Effect of combined waterlogging and salinity stresses on euhalophyte *Suaeda glauca*. Plant Physiol Biochem 127:231–237. https://doi.org/10.1016/j.plaphy.2018.03.030

Duan W, Lu B, Liu L, Meng Y, Ma X, Li J, Zhang K, Sun H, Zhang Y, Dong H, Bai Z, Li C (2022) Effects of exogenous melatonin on root physiology, transcriptome and metabolome of cotton seedlings under salt stress. Int J Mol Sci 23(16):9456. https://doi.org/10.3390/ijms23169456

Eisa EA, Honfi P, Tilly-Mándy A, Mirmazloum I (2023) Exogenous melatonin application induced morpho-physiological and biochemical regulations conferring salt tolerance in *Ranunculus asiaticus* L. Horticulturae 9(2):228. https://doi.org/10.3390/horticulturae9020228

El-Beltagi HS, El-Yazied AA, El-Gawad HGA, Kandeel M, Shalaby TA, Mansour AT, Ibrahim (2023) Synergistic impact of melatonin and putrescine interaction in mitigating salinity stress in snap bean seedlings: reduction of oxidative damage and inhibition of polyamine catabolism. Horticulturae 9(2):285

Erland LAE, Murch SJ, Reiter RJ, Saxena PK (2015) A new balancing act: the many roles of melatonin and serotonin in plant growth and development. Plant Signal Behav 10(11): e1096469. https://doi.org/10.1080/15592324.2015.1096469

Firenzuo A, Vanni P, Ramponi G, Baccari V (1968) Changes in enzyme levels during germination of seeds of *Triticum durum*. Plant Physiol 43:260. https://doi.org/10.1104/pp.43.2.260

Fu J, Wu Y, Miao Y, Xu Y, Zhao E, Wang J, Sun H, Liu Q, Xue Y, Xu Y, Hu T (2017) Improved cold tolerance in *Elymus nutans* by exogenous application of melatonin may involve ABA-dependent and ABA-independent pathways. Sci Rep 7:39865. https://doi.org/10.1038/srep39865

Fukuda A, Nakamura A, Hara N, Toki S, Tanaka Y (2011) Molecular and functional analyses of rice NHXtype Na+/H+ antiporter genes. Planta 233:175–188

Guo J, Li Y, Han G, Song J, Wang B (2018a) NaCl markedly improved the reproductive capacity of the euhalophyte *Suaeda salsa*. Funct Plant Biol 45(3):350–361. https://doi.org/10.1071/FP17181

Guo YY, Tian SS, Liu SS, Wang WQ, Sui N (2018b) Energy dissipation and antioxidant enzyme system protect photosystem II of sweet sorghum under drought stress. Photosynthetica 56(3): 861–872. https://doi.org/10.1007/s11099-017-0741-0

Hao L, Jingjing C, Hejie C, Zhongyuan W, Xiurong G, Chunhua W, Yong Z, Jianxiang M, Jianqiang Y, Xian Z (2017) Exogenous melatonin confers salt stress tolerance to watermelon by improving photosynthesis and redox homeostasis. Plant Sci 8:295

Hardeland R (2016) Melatonin in plants—diversity of levels and multiplicity of functions. Front Plant Sci 7:198. https://doi.org/10.3389/fpls.2016.00198

Hasanuzzaman M, Bhuyan M, Nahar K, Hossain M, Mahmud J, Hossen M, Masud A, Moumita FM (2018) Potassium: a vital regulator of plant responses and tolerance to abiotic stresses. Agronomy (Basel) 8(3):31. https://doi.org/10.3390/agronomy8030031

Hasegawa PM, Bressan RA, Zhu J-K, Bohnert HJ (2000) Plant cellular and molecular responses to high salinity. Annu Rev Plant Physiol Plant Mol Biol 51(1):463–499. https://doi.org/10.1146/annurev.arplant.51.1.463

Hernández-Ruiz J, Arnao MB (2008a) Distribution of melatonin in different zones of lupin and barley plants at different ages in the presence and absence of light. J Agric Food Chem 56(22): 10567–10573. https://doi.org/10.1021/jf8022063

Hernández-Ruiz J, Arnao MB (2008b) Melatonin stimulates the expansion of etiolated lupin cotyledons. Plant Growth Regul 55(1):29–34. https://doi.org/10.1007/s10725-008-9254-y

Hernández-Ruiz J, Cano A, Arnao MB (2004) Melatonin: a growth-stimulating compound present in lupin tissues. Planta 220(1):140–144. https://doi.org/10.1007/s00425-004-1317-3

Hernández-Ruiz J, Cano A, Arnao MB (2005) Melatonin acts as a growth-stimulating compound in some monocot species: melatonin as growth promoter in plants. J Pineal Res 39(2):137–142. https://doi.org/10.1111/j.1600-079X.2005.00226.x

Hu C-H, Zheng Y, Tong C-L, Zhang D-J (2022) Effects of exogenous melatonin on plant growth, root hormones and photosynthetic characteristics of trifoliate orange subjected to salt stress. Plant Growth Regul 97(3):551–558. https://doi.org/10.1007/s10725-022-00814-z

Huang Y-H, Liu S-J, Yuan S, Guan C, Tian D-Y, Cui X, Zhang Y-W, Yang F-Y (2017) Overexpression of ovine *AANAT* and *HIOMT* genes in switchgrass leads to improved growth performance and salt-tolerance. Sci Rep 7(1):12212. https://doi.org/10.1038/s41598-017-12566-2

Huangfu L, Zhang Z, Zhou Y, Zhang E, Chen R, Fang H, Li P, Xu Y, Yao Y, Zhu M, Yin S, Xu C, Lu Y, Yang Z (2021) Integrated physiological, metabolomic and transcriptomic analyses provide insights into the roles of exogenous melatonin in promoting rice seed germination under salt stress. Plant Growth Regul 95(1):19–31. https://doi.org/10.1007/s10725-021-00721-9

Imran M, Shazad R, Bilal S, Imran QM, Khan M, Kang S-M, Khan AL, Yun B-W, Lee I-J (2021) Exogenous melatonin mediates the regulation of endogenous nitric oxide in *Glycine max* L. to reduce effects of drought stress. Environ Exp Bot 188(104511):104511. https://doi.org/10.1016/j.envexpbot.2021.104511

Jabeen N, Ahmad R (2013) The activity of antioxidant enzymes in response to salt stress in safflower (*Carthamus tinctorius* L.) and sunflower (*Helianthus annuus* L.) seedlings raised from seed treated with chitosan: enzyme activity in seedlings from chitosan-treated seed. J Sci Food Agric 93(7):1699–1705. https://doi.org/10.1002/jsfa.5953

Jalili S, Ehsanpour AA, Javadirad SM (2022) The role of melatonin on caspase-3-like activity and expression of the genes involved in programmed cell death (PCD) induced by in vitro salt stress in alfalfa (*Medicago sativa* L.) roots. Bot Stud 63(1):19. https://doi.org/10.1186/s40529-022-00348-7

Jamil M, Rehman S, Rha ES (2007a) Salinity effect on plant growth, PSII photochemistry and chlorophyll content in sugar beet (*Beta vulgaris* L.) and cabbage (*Brassica oleracea capitata* L.). Pak J Bot 39(3):753–760

Jamil M, Rehman S, Lee KJ, Kim JM, Kim H-S, Rha ES (2007b) Salinity reduced growth PS2 photochemistry and chlorophyll content in radish. Sci Agric 64(2):111–118. https://doi.org/10.1590/s0103-90162007000200002

Javeed HMR, Ali M, Skalicky M, Nawaz F, Qamar R, Rehman AU, Faheem M, Mubeen M, Iqbal MM, Rahman MHU, Vachova P, Brestic M, Baazeem A, El Sabagh A (2021) Lipoic acid combined with melatonin mitigates oxidative stress and promotes root formation and growth in salt-stressed canola seedlings (*Brassica napus* L.). Molecules 26(11):3147. https://doi.org/10.3390/molecules26113147

Jiang X, Li H, Song X (2016) Seed priming with melatonin effects on seed germination and seedling growth in maize under salinity stress. Pak J Bot 48(4):1345–1352

Kang K, Lee K, Park S, Kim YS, Back K (2010) Enhanced production of melatonin by ectopic overexpression of human serotonin N-acetyltransferase plays a role in cold resistance in transgenic rice seedlings: transgenic rice plants overproducing melatonin. J Pineal Res 49(2):176–182. https://doi.org/10.1111/j.1600-079X.2010.00783.x

Ke Q, Ye J, Wang B, Ren J, Yin L, Deng X, Wang S (2018) Melatonin mitigates salt stress in wheat seedlings by modulating polyamine metabolism. Front Plant Sci 9:914. https://doi.org/10.3389/fpls.2018.00914

Khan TA, Saleem M, Fariduddin Q (2022) Melatonin influences stomatal behavior, root morphology, cell viability, photosynthetic responses, fruit yield, and fruit quality of tomato plants exposed to salt stress. J Plant Growth Regul 42:1–25

Kolár J, Machá˘cková I, Eder J, Prinsen E, Van Dongen W, Van Onckelen H, Illnerová H (1997) Melatonin: occurrence and daily rhythm in *Chenopodium rubrum*. Phytochemistry 44:1407–1413

Kolár J, Johnson CH, Machá˘cková I (2003) Exogenously applied melatonin (N-acetyl-5-methoxytryptamine) affects flowering of the short-day plant *Chenopodium rubrum*. Physiol Plant 118:605–612

Kołodziejczyk I, Dzitko K, Szewczyk R, Posmyk MM (2016a) Exogenous melatonin improves corn (*Zea mays* L.) embryo proteome in seeds subjected to chilling stress. J Plant Physiol 193:47–56. https://doi.org/10.1016/j.jplph.2016.01.012

Lazár D, Murch SJ, Beilby MJ, Al Khazaaly S (2013) Exogenous melatonin affects photosynthesis in characeae *Chara australis*. Plant Signal Behav 8(3):e23279. https://doi.org/10.4161/psb.23279

Lee H-J, Back K (2016) 2-Hydroxymelatonin promotes the resistance of rice plant to multiple simultaneous abiotic stresses (combined cold and drought). J Pineal Res 61(3):303–316. https://doi.org/10.1111/jpi.12347

Lerner AB, Case JD, Takahashi Y, Lee TH, Mori W (1958) Isolation of melatonin, the pineal gland factor that lightens melanocytes1. J Am Chem Soc 80(10):2587–2587. https://doi.org/10.1021/ja01543a060

Li C, Wang P, Wei Z, Liang D, Liu C, Yin L, Jia D, Fu M, Ma F (2012a) The mitigation effects of exogenous melatonin on salinity-induced stress in *Malus hupehensis*: melatonin alleviates salt stress. J Pineal Res 53(3):298–306. https://doi.org/10.1111/j.1600-079X.2012.00999.x

Li T, Liu R-J, He X-H, Wang B-S (2012b) Enhancement of superoxide dismutase and catalase activities and salt tolerance of euhalophyte *Suaeda salsa* L. by mycorrhizal fungus Glomus mosseae. Pedosphere 22(2):217–224. https://doi.org/10.1016/s1002-0160(12)60008-3

Li W, Yamaguchi S, Khan MA, An P, Liu X, Tran L-SP (2015) Roles of gibberellins and abscisic acid in regulating germination of *Suaeda salsa* dimorphic seeds under salt stress. Front Plant Sci 6:1235. https://doi.org/10.3389/fpls.2015.01235

Li H, Chang J, Chen H, Wang Z, Gu X, Wei C, Zhang Y, Ma J, Yang J, Zhang X (2017a) Exogenous melatonin confers salt stress tolerance to watermelon by improving photosynthesis and redox homeostasis. Front Plant Sci 8:295. https://doi.org/10.3389/fpls.2017.00295

Li X, Yu B, Cui Y, Yin Y (2017b) Melatonin application confers enhanced salt tolerance by regulating Na^+ and Cl^- accumulation in rice. Plant Growth Regul 83(3):441–454. https://doi.org/10.1007/s10725-017-0310-3

Liang C, Zheng G, Li W, Wang Y, Hu B, Wang H, Wu H, Qian Y, Zhu X-G, Tan D-X, Chen S-Y, Chu C (2015) Melatonin delays leaf senescence and enhances salt stress tolerance in rice. J Pineal Res 59(1):91–101. https://doi.org/10.1111/jpi.12243

Liang D, Ni Z, Xia H, Xie Y, Lv X, Wang J, Lin L, Deng Q, Luo X (2019) Exogenous melatonin promotes biomass accumulation and photosynthesis of kiwifruit seedlings under drought stress. Sci Hortic 246:34–43. https://doi.org/10.1016/j.scienta.2018.10.058

Lin J, Li JP, Yuan F, Yang Z, Wang BS, Chen M (2018) Transcriptome profiling of genes involved in photosynthesis in *Elaeagnus angustifolia* L. under salt stress. Photosynthetica 56(4): 998–1009. https://doi.org/10.1007/s11099-018-0824-6

Liu D-D, Sun X-S, Liu L, Shi H-D, Chen S-Y, Zhao D-K (2019) Overexpression of the melatonin synthesis-related gene *SlCOMT1* improves the resistance of tomato to salt stress. Molecules 24(8):1514. https://doi.org/10.3390/molecules24081514

Liu X, Wei Z, Gao T, Zhang Z, Tan K, Li C, Ma F (2021) Ectopic expression of *AANAT* or *HIOMT* improves melatonin production and enhances UV-B tolerance in transgenic apple plants. Fruit Res 1(1):1–13. https://doi.org/10.48130/frures-2021-0004

Mahmoudi H, Massoud RB, Baatour O, Tarchoune I, Salah IB, Nasri N, Abidi W, Kaddour R, Hannoufa A, Lachaâl M, Ouerghi Z (2012) Influence of different seed priming methods for improving salt stress tolerance in lettuce plants. J Plant Nutr 35(12):1910–1922. https://doi.org/10.1080/01904167.2012.711410

Martinez V, Nieves-Cordones M, Lopez-Delacalle M, Rodenas R, Mestre T, Garcia-Sanchez F, Rubio F, Nortes P, Mittler R, Rivero R (2018) Tolerance to stress combination in tomato plants: new insights in the protective role of melatonin. Molecules 23(3):535. https://doi.org/10.3390/molecules23030535

Meloni DA, Oliva MA, Martinez CA, Cambraia J (2003) Photosynthesis and activity of superoxide dismutase, peroxidase and glutathione reductase in cotton under salt stress. Environ Exp Bot 49(1):69–76. https://doi.org/10.1016/s0098-8472(02)00058-8

Mishima K (2012) Melatonin as a regulator of human sleep and circadian systems. Nihon Rinsho 70(7):1139–1144

Mittler R (2002) Oxidative stress, antioxidants and stress tolerance. Trends Plant Sci 7(9):405–410. https://doi.org/10.1016/s1360-1385(02)02312-9
Mukherjee S, David A, Yadav S, Baluška F, Bhatla SC (2014) Salt stress-induced seedling growth inhibition coincides with differential distribution of serotonin and melatonin in sunflower seedling roots and cotyledons. Physiol Plant 152(4):714–728. https://doi.org/10.1111/ppl.12218
Munns R, Tester M (2008) Mechanisms of salinity tolerance. Annu Rev Plant Biol 59(1):651–681. https://doi.org/10.1146/annurev.arplant.59.032607.092911
Murch SJ, KrishnaRaj S, Saxena PK (2000) Tryptophan is a precursor for melatonin and serotonin biosynthesis in in vitro regenerated St. John's wort (*Hypericum perforatum* L. cv. Anthos) plants. Plant Cell Rep 19(7):698–704. https://doi.org/10.1007/s002990000206
Nabati J, Kafi M, Nezami A, Moghaddam PR, Ali M, Mehrjerdi MZ (2011) Effect of salinity on biomass production and activities of some key enzymatic antioxidants in kochia (*Kochia scoparia*). Pak J Bot 43(1):539–548
Olcese J (2000) Melatonin after four decades: an assessment of its potential 2000. Springer, New York, NY
Ouhibi C, Attia H, Rebah F, Msilini N, Chebbi M, Aarrouf J, Urban L, Lachaal M (2014) Salt stress mitigation by seed priming with UV-C in lettuce plants: growth, antioxidant activity and phenolic compounds. Plant Physiol Biochem 83:126–133. https://doi.org/10.1016/j.plaphy.2014.07.019
Padan E, Venturi M, Gerchman Y, Dover N (2001) Na+/H+ antiporters. Biochimica et biophysica acta (BBA)-Bioenergetics 1505(1):144–157
Parida AK, Das AB (2005) Salt tolerance and salinity effects on plants: a review. Ecotoxicol Environ Saf 60(3):324–349. https://doi.org/10.1016/j.ecoenv.2004.06.010
Pelagio-Flores R, Muñoz-Parra E, Ortiz-Castro R, López-Bucio J (2012) Melatonin regulates *Arabidopsis* root system architecture likely acting independently of auxin signaling: role of melatonin in root architecture and auxin responses. J Pineal Res 53(3):279–288. https://doi.org/10.1111/j.1600-079X.2012.00996.x
Pitman MG, Läuchli A (2006) Global impact of salinity and agricultural ecosystems. In: Salinity: environment—plants—molecules. Kluwer Academic, Dordrecht, pp 3–20
Posmyk MM, Janas KM (2009) Melatonin in plants. Acta Physiol Plant 31(1):1–11. https://doi.org/10.1007/s11738-008-0213-z
Qi C, Zhang H, Liu Y, Wang X, Dong D, Yuan X, Li X, Zhang X, Li X, Zhang N, Guo Y-D (2020) *CsSNAT* positively regulates salt tolerance and growth of cucumber by promoting melatonin biosynthesis. Environ Exp Bot 175:104036. https://doi.org/10.1016/j.envexpbot.2020.104036
Rahnama A, James RA, Poustini K, Munns R (2010) Stomatal conductance as a screen for osmotic stress tolerance in durum wheat growing in saline soil. Funct Plant Biol 37(3):255. https://doi.org/10.1071/fp09148
Raza SH, Athar HUR, Ashraf M (2006) Influence of exogenously applied glycinebetaine on the photosynthetic capacity of two differently adapted wheat cultivars under salt stress. Pak J Bot 38(2):341–351
Reiter RJ (1991) Pineal melatonin: cell biology of its synthesis and of its physiological interactions. Endocr Rev 12(2):151–180. https://doi.org/10.1210/edrv-12-2-151
Ren Z, Shi J, Guo A, Wang Y, Fan X, Li R, Yu C, Peng Z, Gao Y, Liu Z, Duan L (2022) Melatonin mediates the regulation of morphological and anatomical traits in *Carex leucochlora* under continuous salt stress. Agronomy (Basel) 12(9):2098. https://doi.org/10.3390/agronomy12092098
Rodriguez C, Mayo JC, Sainz RM, Antolín I, Herrera F, Martín V, Reiter RJ (2004) Regulation of antioxidant enzymes: a significant role for melatonin. J Pineal Res 36(1):1–9. https://doi.org/10.1046/j.1600-079x.2003.00092.x
Sairam RK, Srivastava GC (2002) Changes in antioxidant activity in sub-cellular fractions of tolerant and susceptible wheat genotypes in response to long term salt stress. Plant Sci 162(6):897–904. https://doi.org/10.1016/s0168-9452(02)00037-7

Sardar H, Ramzan MA, Naz S, Ali S, Ejaz S, Ahmad R, Altaf MA (2023) Exogenous application of melatonin improves the growth and productivity of two broccoli (*Brassica oleracea* L.) cultivars under salt stress. J Plant Growth Regul 42:1–15

Shalata A, Mittova V, Volokita M, Guy M, Tal M (2001) Response of the cultivated tomato and its wild salt-tolerant relative *Lycopersicon pennellii* to salt-dependent oxidative stress: the root antioxidative system. Physiol Plant 112(4):487–494. https://doi.org/10.1034/j.1399-3054.2001.1120405.x

Sharma A, Zheng B (2019) Melatonin mediated regulation of drought stress: physiological and molecular aspects. Plan Theory 8(7):190

Sharma A, Wang J, Xu D, Tao S, Chong S, Yan D, Zheng B et al (2020) Melatonin regulates the functional components of photosynthesis, antioxidant system, gene expression, and metabolic pathways to induce drought resistance in grafted *Carya cathayensis* plants. Sci Total Environ 713:136675

Shi H, Zhu J-K (2002) Regulation of expression of the vacuolar Na^+/H^+ antiporter gene AtNHX1 by salt stress and abscisic acid. Plant Mol Biol 50(3):543–550

Shi H, Jiang C, Ye T, Tan D-X, Reiter RJ, Zhang H, Liu R, Chan Z (2015a) Comparative physiological, metabolomic, and transcriptomic analyses reveal mechanisms of improved abiotic stress resistance in bermudagrass [*Cynodon dactylon* (L). Pers.] by exogenous melatonin. J Exp Bot 66(3):681–694. https://doi.org/10.1093/jxb/eru373

Shi H, Qian Y, Tan D-X, Reiter RJ, He C (2015b) Melatonin induces the transcripts of *CBF/DREB1*s and their involvement in both abiotic and biotic stresses in *Arabidopsis*. J Pineal Res 59(3):334–342. https://doi.org/10.1111/jpi.12262

Shi H, Reiter RJ, Tan D-X, Chan Z (2015c) Indole-3-acetic acid inducible 17 positively modulates natural leaf senescence through melatonin-mediated pathway in *Arabidopsis*. J Pineal Res 58(1):26–33. https://doi.org/10.1111/jpi.12188

Shi H, Wang X, Tan D-X, Reiter RJ, Chan Z (2015d) Comparative physiological and proteomic analyses reveal the actions of melatonin in the reduction of oxidative stress in Bermuda grass (*Cynodon dactylon* (L). Pers.). J Pineal Res 59(1):120–131. https://doi.org/10.1111/jpi.12246

Simlat M, Ptak A, Wójtowicz T, Szewczyk A (2023) The content of phenolic compounds in *Stevia rebaudiana* (Bertoni) plants derived from melatonin and NaCl treated seeds. Plan Theory 12(4):780. https://doi.org/10.3390/plants12040780

Sun W, Van Montagu M, Verbruggen N (2002) Small heat shock proteins and stress tolerance in plants. Biochim Biophys Acta Gene Struct Expr 1577(1):1–9. https://doi.org/10.1016/s0167-4781(02)00417-7

Sun S, Wen D, Yang W, Meng Q, Shi Q, Gong B (2020) Overexpression of caffeic acid O-methyltransferase 1 (COMT1) increases melatonin level and salt stress tolerance in tomato plant. J Plant Growth Regul 39(3):1221–1235. https://doi.org/10.1007/s00344-019-10058-3

Takahashi S, Murata N (2008) How do environmental stresses accelerate photo inhibition? Trends Plant Sci 13:178–182

Tal O, Haim A, Harel O, Gerchman Y (2011) Melatonin as an antioxidant and its semi-lunar rhythm in green macroalga *Ulva* sp. J Exp Bot 62(6):1903–1910. https://doi.org/10.1093/jxb/erq378

Talaat NB, Shawky BT (2022) Synergistic effects of salicylic acid and melatonin on modulating ion homeostasis in salt-stressed wheat (*Triticum aestivum* L.) plants by enhancing root H^+-pump activity. Plan Theory 11(3):416. https://doi.org/10.3390/plants11030416

Talaat NB, Todorova D (2022) Antioxidant machinery and glyoxalase system regulation confers salt stress tolerance to wheat (*Triticum aestivum* L.) plants treated with melatonin and salicylic acid. J Soil Sci Plant Nutr 22(3):3527–3540. https://doi.org/10.1007/s42729-022-00907-8

Tan D-X, Reiter RJ, Manchester LC, Yan M-T, El-Sawi M, Sainz RM, Mayo JC, Kohen R, Allegra M, Hardeland R (2002) Chemical and physical properties and potential mechanisms: melatonin as a broad spectrum antioxidant and free radical scavenger. Curr Top Med Chem 2(2):181–197. https://doi.org/10.2174/1568026023394443

Tan D-X, Manchester LC, Liu X, Rosales-Corral SA, Acuna-Castroviejo D, Reiter RJ (2013a) Mitochondria and chloroplasts as the original sites of melatonin synthesis: a hypothesis related to melatonin's primary function and evolution in eukaryotes. J Pineal Res 54(2):127–138. https://doi.org/10.1111/jpi.12026

Tan HW, Abdul Aziz AR, Aroua MK (2013b) Glycerol production and its applications as a raw material: a review. Renew Sustain Energy Rev 27:118–127. https://doi.org/10.1016/j.rser.2013.06.035

Tan D-X, Hardeland R, Back K, Manchester LC, Alatorre-Jimenez MA, Reiter RJ (2016) On the significance of an alternate pathway of melatonin synthesis via 5-methoxytryptamine: comparisons across species. J Pineal Res 61(1):27–40. https://doi.org/10.1111/jpi.12336

Tan K, Zheng J, Liu C, Liu X, Liu X, Gao T, Song X, Wei Z, Ma F, Li C (2021) Heterologous expression of the melatonin-related gene *HIOMT* improves salt tolerance in *Malus domestica*. Int J Mol Sci 22(22):12425. https://doi.org/10.3390/ijms222212425

Taspinar MS, Kukara M, Sigmaz B, Agar G (2023) The effect of melatonin on some physiological parameters, salt sensor genes and DNA methylation in salt stressed lettuce (*Lactuca Sativa* L.) Author preprints. https://doi.org/10.22541/au.167603127.71746463/v1

Timperio AM, Egidi MG, Zolla L (2008) Proteomics applied on plant abiotic stresses: role of heat shock proteins (HSP). J Proteomics 71(4):391–411. https://doi.org/10.1016/j.jprot.2008.07.005

Ulfat M, Athar HR, Ashraf M, Akram NA (2007) Appraisal of physiological and biochemical selection criteria for evaluation of salt tolerance in canola (*Brassica napus* L.). Pak J Bot 39(5): 1593–1608

Van Tassel DL, Roberts N, Lewy A, O'Neill SD (2001) Melatonin in plant organs. J Pineal Res 31(1):8–15. https://doi.org/10.1034/j.1600-079x.2001.310102.x

Vierling E (1991) The roles of heat-shock proteins in plants. Annu Rev Plant Physiol Plant Mol Biol 42:579–620. https://doi.org/10.1146/annurev.arplant.42.1.579

Wang J, Liu D, Guo X, Zhang A (2012a) Research progress in the pathway of auxin synthesis. Acta Bot Sin 47:292–301

Wang P, Yin L, Liang D, Li C, Ma F, Yue Z (2012b) Delayed senescence of apple leaves by exogenous melatonin treatment: toward regulating the ascorbate-glutathione cycle: melatonin regulates the AsA-GSH cycle. J Pineal Res 53(1):11–20. https://doi.org/10.1111/j.1600-079X.2011.00966.x

Wang P, Sun X, Xie Y, Li M, Chen W, Zhang S, Liang D, Ma F (2014) Melatonin regulates proteomic changes during leaf senescence in *Malus hupehensis*. J Pineal Res 57(3):291–307. https://doi.org/10.1111/jpi.12169

Wang LY, Liu JL, Wang WX, Sun Y (2016a) Exogenous melatonin improves growth and photosynthetic capacity of cucumber under salinity-induced stress. Photosynthetica 54(1): 19–27. https://doi.org/10.1007/s11099-015-0140-3

Wang Q, An B, Wei Y, Reiter RJ, Shi H, Luo H, He C (2016b) Melatonin regulates root meristem by repressing auxin synthesis and polar auxin transport in *Arabidopsis*. Front Plant Sci 7:1882. https://doi.org/10.3389/fpls.2016.01882

Wang D, Hao Z, Zhao J, Jin Y, Huang J, Zhou R, Wu C (2019) Comparative physiological and transcriptomic analyses reveal salt tolerance mechanisms of *Zygosaccharomyces rouxii*. Process Biochem 82:59–67. https://doi.org/10.1016/j.procbio.2019.04.009

Wang J, Lv P, Yan D, Zhang Z, Xu X, Wang T, Wang Y, Peng Z, Yu C, Gao Y, Duan L, Li R (2022) Exogenous melatonin improves seed germination of wheat (*Triticum aestivum* L.) under salt stress. Int J Mol Sci 23(15):8436. https://doi.org/10.3390/ijms23158436

Wang F, Wan C, Wu W, Yang S, Chen X (2023) Exogenous melatonin (MT) enhances salt tolerance of okra (*Abelmoschus esculentus* L.) plants by regulating proline, photosynthesis, ion homeostasis and ROS pathways. Vegetos:1–15

Wei W, Li Q-T, Chu Y-N, Reiter RJ, Yu X-M, Zhu D-H, Zhang W-K, Ma B, Lin Q, Zhang J-S, Chen S-Y (2015) Melatonin enhances plant growth and abiotic stress tolerance in soybean plants. J Exp Bot 66(3):695–707. https://doi.org/10.1093/jxb/eru392

Wei Y, Liu G, Chang Y, Lin D, Reiter RJ, He C, Shi H (2018) Melatonin biosynthesis enzymes recruit *WRKY* transcription factors to regulate melatonin accumulation and transcriptional activity on W-box in cassava. J Pineal Res 65(1):e12487. https://doi.org/10.1111/jpi.12487

Wei J, Liang J, Liu D, Liu Y, Liu G, Wei S (2022a) Melatonin-induced physiology and transcriptome changes in banana seedlings under salt stress conditions. Front Plant Sci 13: 938262. https://doi.org/10.3389/fpls.2022.938262

Wei L, Zhao H, Wang B, Wu X, Lan R, Huang X, Chen B, Chen G, Jiang C, Wang J, Liu Y, Zheng Q (2022b) Exogenous melatonin improves the growth of rice seedlings by regulating redox balance and ion homeostasis under salt stress. J Plant Growth Regul 41(6):2108–2121. https://doi.org/10.1007/s00344-021-10417-z

Wu C, Cao S, Xie K, Chi Z, Wang J, Wang H, Wei Y, Shao X, Zhang C, Xu F, Gao H (2021) Melatonin delays yellowing of broccoli during storage by regulating chlorophyll catabolism and maintaining chloroplast ultrastructure. Postharv Biol Technol 172:111378. https://doi.org/10.1016/j.postharvbio.2020.111378

Xu H-H, Liu S-J, Song S-H, Wang R-X, Wang W-Q, Song S-Q (2016a) Proteomics analysis reveals distinct involvement of embryo and endosperm proteins during seed germination in dormant and non-dormant rice seeds. Plant Physiol Biochem 103:219–242. https://doi.org/10.1016/j.plaphy.2016.03.007

Xu Y-G, Liu R, Sui N, Shi W, Wang L, Tian C, Song J (2016b) Changes in endogenous hormones and seed-coat phenolics during seed storage of two *Suaeda salsa* populations. Aust J Bot 64(4): 325. https://doi.org/10.1071/bt16014

Yandi W (2018) Chinese changes in melatonin content in grape and prokaryotic expression analysis of its synthetic gene SNAT. Academy of Agricultural Sciences. Academy of Agricultural Sciences, Beijing

Ye J, Wang S, Deng X, Yin L, Xiong B, Wang X (2016) Melatonin increased maize (*Zea mays* L.) seedling drought tolerance by alleviating drought-induced photosynthetic inhibition and oxidative damage. Acta Physiol Plant 38(2):48. https://doi.org/10.1007/s11738-015-2045-y

Yu HS, Hernandez V, Haywood M, Wong CG (1993) Melatonin inhibits the proliferation of retinal pigment epithelial (RPE) cells in vitro. In Vitro Cell Dev Biol Anim 29A(5):415–418. https://doi.org/10.1007/bf02633991

Zhan Y, Wu T, Zhao X, Wang Z, Chen Y (2021) Comparative physiological and full-length transcriptome analyses reveal the molecular mechanism of melatonin-mediated salt tolerance in okra (*Abelmoschus esculentus* L.). BMC Plant Biol 21(1):180. https://doi.org/10.1186/s12870-021-02957-z

Zhang N, Zhao B, Zhang H-J, Weeda S, Yang C, Yang Z-C, Ren S, Guo Y-D (2013) Melatonin promotes water-stress tolerance, lateral root formation, and seed germination in cucumber (*Cucumis sativus* L.): melatonin alleviates water stress in cucumber. J Pineal Res 54(1): 15–23. https://doi.org/10.1111/j.1600-079X.2012.01015.x

Zhang H-J, Zhang N, Yang R-C, Wang L, Sun Q-Q, Li D-B, Cao Y-Y, Weeda S, Zhao B, Ren S, Guo Y-D (2014a) Melatonin promotes seed germination under high salinity by regulating antioxidant systems, ABA and GA4 interaction in cucumber (*Cucumis sativus* L.). J Pineal Res 57(3):269–279. https://doi.org/10.1111/jpi.12167

Zhang N, Zhang H-J, Zhao B, Sun Q-Q, Cao Y-Y, Li R, Wu X-X, Weeda S, Li L, Ren S, Reiter RJ, Guo Y-D (2014b) The RNA-seq approach to discriminate gene expression profiles in response to melatonin on cucumber lateral root formation. J Pineal Res 56(1):39–50. https://doi.org/10.1111/jpi.12095

Zhang N, Zhang H-J, Sun Q-Q, Cao Y-Y, Li X, Zhao B, Wu P, Guo Y-D (2017) Proteomic analysis reveals a role of melatonin in promoting cucumber seed germination under high salinity by regulating energy production. Sci Rep 7(1):503. https://doi.org/10.1038/s41598-017-00566-1

Zhang C, Xu B, Zhao C-R, Sun J, Lai Q, Yu C (2019) Comparative de novo transcriptomics and untargeted metabolomic analyses elucidate complicated mechanisms regulating celery (*Apium graveolens* L.) responses to selenium stimuli. PloS One 14(12):e0226752. https://doi.org/10.1371/journal.pone.0226752

Zhao K-F, Song J, Fan H, Zhou S, Zhao M (2010) Growth response to ionic and osmotic stress of NaCl in salt-tolerant and salt-sensitive maize. J Integr Plant Biol 52(5):468–475. https://doi.org/10.1111/j.1744-7909.2010.00947.x

Zhu JK (2003) Regulation of ion homeostasis under salt stress. Curr Opin Plant Biol 6(5):441–445. https://doi.org/10.1016/s1369-5266(03)00085-2

Zhu B, Zheng S, Fan W, Zhang M, Xia Z, Chen X, Zhao A (2022) Ectopic overexpression of mulberry *MnT5H2* enhances melatonin production and salt tolerance in tobacco. Front Plant Sci 13:1061141. https://doi.org/10.3389/fpls.2022.1061141

Zhuang W, Liu T, Shu X, Wang H, Wang Z, Wang T, Zhang F, Qu S (2020) Overexpression of *MzASMT 1*, a gene from *Malus zumi* Mats, enhances salt tolerance in transgenic tobacco. Front Plant Sci 11:561903. https://doi.org/10.3389/fpls.2020.561903

Zuo B, Zheng X, He P, Wang L, Lei Q, Feng C, Zhou J, Li Q, Han Z, Kong J (2014) Overexpression of *MzASMT* improves melatonin production and enhances drought tolerance in transgenic *Arabidopsis thaliana* plants. J Pineal Res 57(4):408–417. https://doi.org/10.1111/jpi.12180

Chapter 9
Melatonin and Cold Tolerance in Plants

Jinxia Cui and Pei Wu

Abstract Cold is one of the most adverse environmental factors affecting plant growth and development. Since melatonin was first detected in plants in 1995, more and more studies have reported the mechanisms by which melatonin enhances the cold resistance of plants. In this chapter, we comprehensively summarized the mechanism of melatonin-induced enhancement of plant cold tolerance covering the synthesis and metabolism of melatonin, the crosstalk between melatonin and other small signal molecules, and the impact of melatonin on the expression levels of cold-responsive genes under cold stress. This review advances our current understanding of the mechanism of melatonin in the circumstance of plant response to cold stress.

Keywords Melatonin · Cold stress · Signal transduction · Gene expression

9.1 Introduction

Cold spell plays an indispensable role in the physiology of plant species. Some critical steps of plant development, such as seed germination, flowering, dormancy, and so on, usually require seeds (or plants) to be exposed to cold environments (Auge et al. 2017). Some certain plant species need to promote their flowering process through prolonged exposure to the cold of a typical winter (Amasino 2005). Warne (1947) reported the importance of the low temperature (2–8 °C) treatment on the swollen seeds at different periods, which shortened the period of plants staying in the heading stage and accelerated flowering for 14–20 days without affecting the heading date. In addition, strawberries require a medium or higher

J. Cui (✉) · P. Wu
Department of Horticulture, College of Agriculture, Shihezi University, Shihezi, China

The Key Laboratory of Special Fruits and Vegetables Cultivation Physiology and Germplasm Resources Utilization of Xinjiang Production and Construction Group, Shihezi University, Shihezi, China

A. Sharma, G. J. Ahammed (eds.), *Melatonin in Plants: Role in Plant Growth, Development, and Stress Response*, Plant Life and Environment Dynamics,
https://doi.org/10.1007/978-981-99-8051-2_9

temperature for color transition and sugar synthesis, but a low temperature of about 10 °C is required to form a unique aroma. Fruits grown at about 10 °C have a stronger aroma.

Conversely, cold stress can also impose threats on plant's physiological activity, molecular and biochemical metabolism, growth and development, and crop productivity (Manasa et al. 2021). Plants' adaptation to cold stress is a complex and huge network involving many cellular processes. The cold stress signal is first perceived by the receptors located on the cell membrane, such as transporter proteins, protective proteins, and regulatory proteins. And then, the intracellular signaling cascades including calcium (Ca^{2+}), redox, phytohormone, and mitogen-activated protein kinases (MAPK) signals are activated. Some regulatory proteins can also act as signal transducers and transcriptional activators by undergoing phosphorylation (Ph), glycosylation (G), sumoylation (SUMO), oxidation (Ox), and other modifications in signal cascade reactions. Finally, the signal cascades induced the expression of cold-responsive genes and their protein products can develop cold tolerance directly or indirectly (Janmohammadi et al. 2015; Zhang et al. 2023; Zhao et al. 2017).

Melatonin, a kind of indoleamine, was found in bovine pineal glands (Shafi et al. 2021). So far, melatonin has been measured in many plant species with different tissues including leaves, roots, buds, and fruits (Murch et al. 2000; Paredes et al. 2009), and it is also found to function as a signal molecule in plants' growth, development, and defense response against abiotic stresses (Tan et al. 2015; Hernandez et al. 2015). Under cold stress, melatonin participates in regulating the membrane integrity, stomatal conductance, photosynthetic efficiency, antioxidative defense, redox homeostasis, the accumulation of osmolytes, hormones and secondary metabolites, and the expression of stress-responsive genes (Qari et al. 2022). In this chapter, we summarize the current knowledge of how melatonin is induced in plants when exposed to cold stress and the defensive functions of melatonin in plants regarding cold stress tolerance (including the knowledge at the physiological and molecular levels).

9.2 Effect of Cold Stress on Melatonin Synthesis and Catabolism

In recent years, many studies have reported an increase in the production of melatonin in plants' response to cold stress (Li et al. 2021; Raza et al. 2022) (Table 9.1). Among the many species studied, only rice (*Oryza sativa*) had a decrease in melatonin content when exposed to cold stress (Han et al. 2017). It reported that the tryptophan decarboxylase (TDC), serotonin *N*-acetyltransferase (SNAT), tryptamine 5-hydroxylase (T5H), *N*-acetylserotonin methyltransferase (ASMT), tryptophan hydroxylase (TPH), and caffeic acid *O*-methyltransferase (COMT) were involved in the melatonin synthesis from tryptophan with multiple

Table 9.1 Effect of cold stress on melatonin production

Plant species	Organs/ tissue	Treatments (low temperatures)	Treatment duration	Cold effect	References
Citrullus lanatus	Leaf	4 °C	12 h	+	Li et al. (2021)
Cucumis sativus	Leaf	5 °C	0–24 h	+	Feng et al. (2021)
Solanum lycopersicum	Leaf	4 °C	24 h	+	Ding et al. (2021)
Cynodon dactylon	Leaf	−5–4 °C	8 h–7 days	+	Fan et al. (2015)
Cynodon dactylon	Leaf	4 °C	3 days	+	Hu et al. (2016)
Hordeum vulgare	Leaf	2–4 °C	48 h	+	Li et al. (2016)
Oryza sativa	Leaf	12 °C	6 days	–	Han et al. (2017)
Fragaria ananassa	Leaf	−4–0 °C	2 days	+	Hayat et al. (2022)
Lupinus albus	Root	6 °C	18–36 h	+	Arnao and Hernandez-Ruiz (2013)
Solanum lycopersicum	Fruit	4 ± 0.5 °C	7–28 days	+	Sharafi et al. (2019)
Datura metel	Young flower buds	4 °C	3 days	+	Murch et al. (2009)
Hulless barley	Leaf	5–15 °C	72 h	+	Chang et al. (2021)
Solanum melongena	Fruit	5 ± 0.5 °C	5–15 days	+	Song et al. (2022)
Arabidopsis thaliana	Leaf	4 °C	0–24 h	+	Shi and Chan (2014)
Citrullus lanatus	Leaf/root	4–10 °C	24–72 h	+	Li et al. (2017)

pathways in plants. However, ASMT has been identified to have the greatest impact on melatonin content in plants among these six enzymes (Yang et al. 2022; Back et al. 2016). The intermediates of melatonin can be produced in different subcellular compartments such as cytoplasm, endoplasmic reticulum, and chloroplasts, which may also affect the subsequent enzymatic steps in the process of melatonin synthesis. In addition, different pathways of melatonin synthesis lead to its synthesis at different subcellular sites, such as in mitochondria or chloroplasts, which may also affect the mode of action of melatonin in plants to varying degrees (Back et al. 2016).

Melatonin is often induced by cold and several other environmental stressors such as drought, salinity, and chemical toxins, and melatonin could also further induce the expression of cold and stress-responsive genes in Arabidopsis (Hardeland 2015). The content of melatonin was closely related to the regulation of synthesis and its metabolites, all of which can act as biological stimulators or signaling molecules for a series of biological functions in plants (Arnao and Hernandez-Ruiz 2015; Arnao and Hernndez-Ruiz 2014). Melatonin is useful for improving the survival of long-term storage of naturally resistant elm germplasm through the process of cryopreservation (Uchendu et al. 2013).

9.3 Melatonin and Plant Response to Cold

Cold stress is closely related to plant cold acclimation and development. Fortunately, the role of melatonin has been studied in the context of cold (chilling and freezing) tolerance. Both genetic and pharmacological approaches proved that the production of melatonin is required for plants' cold tolerance (Arnao and Hernandez-Ruiz 2013; Hwang and Back 2019). In this instance, melatonin is essential either for survival under cold stress in non-acclimated plants or for the achievement of further cold tolerance in cold-acclimated plants. A significant increase in melatonin accumulation has been observed in serotonin *N*-acetyltransferase (SNAT)-overexpressed plants, accompanied by an increase in chlorophyll content (Kang et al. 2010). A previous studies have reported that melatonin plays a potential role in regulating combined cold and drought stress, and this indicated that melatonin metabolites participate in plant growth and development as functional substances under abiotic stress (Lei et al. 2013; Sharma and Zheng 2019; Sharma et al. 2020). In addition, melatonin effects as growth promoter and signal molecule mediate plants response to cold stress. The effects are dose-dependent, specifically, alleviating cold damage and provoking root growth at low concentrations, while exacerbating cold damage of cucumber seedlings and repressing the leaf area at high concentrations (Wu et al. 2022; Song et al. 2022; Hernandez et al. 2015).

Melatonin signaling also participates in the regulation of photosynthesis and the tricarboxylic acid cycle in increasing the cold tolerance of plants. Melatonin enhances the carbon assimilation in higher carbon dioxide (CO_2) concentrations by increasing the ATPase activities, sucrose synthesis, net photosynthetic rate (Pn), and the higher expression level of total chlorophyll concentration and biomass accumulation in plants (Zhou et al. 2020). And these can indicate that the application of melatonin during grain filling in maternal plants may be an effective way to improve the cold resistance of wheat offspring at the seedling stage (Li et al. 2018). Under cold stress, melatonin could alter the photosynthetic fluorescence parameters of bermudagrass, in both the cold-sensitive and cold-tolerant cultivars, indicating the positive effect of melatonin in regulating cold stress by improving the photosystem function (Hu et al. 2016). Furthermore, melatonin could accelerate the rate of the tricarboxylic acid cycle, the respiration rate of mitochondrial, and cytochrome pathways, the alternative respiration rate, and the activity of ATP synthase and ATP concentration under cold stress (Turk and Genisel 2020). In addition, some studies reported that the higher chlorophyll content, Pn, photochemical activity of photosystem (PS) II and PSI, stomatal conductance, the maximal quantum yield of PSII photochemistry (Fv/Fm), and the efficiency of excitation energy capture of open PSII centers (TR_O/RC) were observed in melatonin-treated melon plants (Zhang et al. 2017; Altaf et al. 2022).

Melatonin was also found to markedly enhance antioxidant capacity and maintain redox homeostasis in plant response to cold, especially in fruit cold storage. Studies have shown that melatonin can delay senescence and maintain the quality of fresh-cut broccoli florets in cold storage (Wei et al. 2020; Li et al. 2019). And these effects were mainly due to melatonin-induced positive effects on cell membrane lipid peroxidation, permeability, structure integrity, and antioxidant defense system (Sun et al. 2020; Kong et al. 2020). Melatonin could maintain the nutraceutical properties of pomegranate fruits during cold storage by ensuring adequate intracellular NADPH supply, which is attributed to its increased glucose-6-phosphate dehydrogenase (G6PDH) and 6-phosphogluconate dehydrogenase (6PGDH) activity, promoted phenylpropanoid pathway activity, and reactive oxygen species (ROS) prevention and clearance system activity (Aghdam et al. 2020). Melatonin also improved the postharvest quality of eggplants during cold storage by enhancing antioxidant capacity and inhibiting cell wall degradation (Song et al. 2022).

Cell membranes are the first line of defense for plants to receive cold signals and the cell membranes are the most vulnerable part during the cold storage of fruit. Therefore, maintaining the integrity and function of cell membranes is conducive to decreasing the chilling injury of fruits. Fortunately, researchers found that melatonin could remodel the lipid molecules and maintain membrane structure by regulating membrane lipid metabolism, and further activating downstream signaling pathways, and ultimately improving the cold resistance of mango fruits (Xu et al. 2023). The important evidence for melatonin to improve cold tolerance of peach fruit after harvest was to reduce mealiness by positively modifying numerous enzymes and proteins located in the cell wall, and maintaining the depolymerization state of cell wall polysaccharides (Cao et al. 2018). In addition, melatonin could delay the senescence of sweet cherry by reducing respiration rate, pedicel browning, and maintaining higher total soluble solids (TSS) and total soluble solids/titratable acidity (TSS/TA) ratio. Melatonin also alleviated cold damage and induced a higher nutritional quality of fruits by suppressing fruit senescence and preserved energy status. Moreover, melatonin inhibited the cold-elicited "ethylene burst" and transcription factors associated with MYBs, but it increased the anthocyanin levels and the content of bioactive compounds, and regulated the energy and proline metabolism during cold storage (Miranda et al. 2020; Xu et al. 2022; Liu et al. 2020).

The circadian rhythms of the plant were also been found to be regulated by melatonin under cold stress. Melatonin-induced circadian rhythms of *hulless barley* by influencing the expression profiles of clock genes and increasing some key physiological indicators, ultimately enhancing the resistance of *hulless barley* seedlings to cold stress (Chang et al. 2021). At the same time, researchers also reported the relationship between circadian rhythms and the melatonin biosynthesis pathway in rice under abiotic stresses (Ahn et al. 2021).

9.4 Melatonin Signal Transduction Under Cold Stress

Whether or not melatonin can participate in regulating the plant abiotic stress tolerance through plant hormone and secondary signal transduction networks depends on the recognition of molecular signals, high-affinity binding sites, and signal transduction pathways. For a long time, melatonin has been considered as a positive antioxidant, which can alleviate abiotic stress by directly eliminating excess ROS in plants. However, there is increasing evidence indicating that melatonin, as a signal molecule, does not function independently, but interacts with some other signal molecules including Ca^{2+}, nitric oxide (NO), hydrogen peroxide (H_2O_2), brassinosteroid (BR), and other plant hormones such as auxin (IAA), abscisic acid (ABA), jasmonic acid (JA) to transfer the stress signals.

Melatonin and Ca^{2+} synergistically boosted cold tolerance in *Dalbergia odorifera* T. Chen and cucumber seedlings by regulating photosynthetic efficiency, ROS scavenging, redox homeostasis, and maintaining osmotic balance and content of diverse phytohormones (Pu et al. 2021; Ma et al. 2022). Besides, melatonin confers cold tolerance to cucumber seedlings in a Ca^{2+} signaling (Ca^{2+}/CDPKs and Ca^{2+}/CaM)-dependent manner (Ma et al. 2022). Cold-induced accumulation of H_2O_2 may serve as a signaling molecule to increase endogenous melatonin content by regulating the expression level of melatonin biosynthesis genes (*TDC*, *T5H*, *SNAT*, and *ASMT*) under low temperature stress (Sharafi et al. 2019). Researchers also found that melatonin-BRs-H_2O_2 signaling networks enhance the cold resistance of turfgrass plants by regulating redox homeostasis. And in the initial stages of cold stress, exogenous melatonin could increase endogenous melatonin levels and induces H_2O_2-dependent BRs biosynthetic signaling pathway, which might serve as downstream components to further activate transcription factors and genes involved in abiotic stress, including dehydration-responsive element binding protein gene (*DREB1A*), ethylene responsive factor (*ERF109*), and *MYB108* (Fu et al. 2022; Liu et al. 2019; Wang et al. 2019; Hu et al. 2020). In addition, melatonin could notably increase the level of putrescine (Put) and spermidine (Spd), which may be responsible for alleviating cold-induced apoptosis (Lei et al. 2004). However, polyamines can be decomposed by polyamine oxidase (PAO) to produce NO in plants under cold stress (Mur et al. 2013). This report provides an important clue to studying the interaction between melatonin and NO. Melatonin regulated the NO/NOS system through various mechanisms to affect physiological and pathological processes in the human body (Fan et al. 2018). Similarly, melatonin and NO also interact with each other to regulate plant growth and development and responses to environmental stress (He and He 2020). A recent study showed that exogenous melatonin triggered arginine-dependent NO accumulation and conferred tomato fruits with chilling resistance by regulating cold-related genes (Aghdam et al. 2019).

Crosstalk between melatonin and other plant hormones is also involved in regulating plant cold adaptive responses. A study reported that both melatonin and BR levels were simultaneously decreased by suppressing *SNAT2* gene, indicating an interactive relationship between melatonin and BR (Hwang and Back 2019).

Exogenous melatonin induced an increase in the content of IAA and gibberellic acid (GA3), while a decrease in ABA accumulation during plant response to cold stress, indicating the crosstalk between melatonin and other phytohormones (Pu et al. 2021). A previous study reported that melatonin might function synergistically with JA and IAA, while antagonistically with ABA in regulating the responses of plants to cold stress. And melatonin-induced accumulation of JA activates the CBF-response pathway, which further enhances the redox state and photosynthetic capacity of plants. In addition, JA-induced accumulation of IAA could also stabilize plant redox levels by regulating photosynthesis to improve plant low temperature tolerance (Chang et al. 2020).

9.5 Melatonin-Induced Gene Expression Under Cold Stress

Many studies have indicated that melatonin plays a crucial role in controlling cold-responsive gene transcription (Qari et al. 2022). Recently, a study revealed that melatonin improved cold resistance of *Arabidopsis thaliana* through regulating expression levels of several cold-related genes, such as C-repeat binding factors/ drought response element binding factors (*CBFs/DREBs*), *COR15a*, calmodulin-binding transcription activator (*CAMTA1*), zinc finger of arabidopsis thaliana 10 (*ZAT10*), and *ZAT12* (Bajwa et al. 2014). In addition, *ZAT6* was also involved in melatonin-mediated freezing stress response in Arabidopsis, and the expression level of *AtZAT6* significantly affected the induction *AtCBFs* transcripts by exogenous melatonin treatment under cold stress (Shi and Chan 2014). Subsequently, Li et al. (2017) reported that melatonin can also affect the expression levels of some genes, which are involved in signaling cascades, such as calcium-dependent protein kinase 18 (*CDPK18*), mitogen-activated protein kinase 16 (*MAPK16*), respiratory burst oxidase homolog (*RBOH*), and *RBOH-like* and transcription factors including ethylene responsive transcription factor (*ERF*), *MYB-like*, basic helix-loop-helix (*bHLH*), basic leucine zipper domain (*bZIP*), heat stress transcription factor (*HSF*), and *WRKY* (Li et al. 2017). Previous studies have shown that melatonin participates in regulating plant cold resistance by regulating photosynthetic carbon assimilation efficiency. Furthermore, Ding et al. (2017) found that exogenous melatonin significantly promotes the expression levels of cold response genes, including *SlICE*, *SlCBF*, *SlP5CS,* and *SlSBP*. It is noteworthy that *SlSBP* encodes the enzyme of sedoheptulose 1,7-diphosphatase (SBPase), which is involved in the Calvin cycle, and is significantly induced in plants after treatment with melatonin under cold stress, these results are consistent with the observed increase in photosynthetic carbon assimilation.

9.6 Conclusions

Melatonin, as a potent regulator of plant growth and stress tolerance, could be induced by cold in multiple tissues of plants such as leaves, roots, buds, and fruits through diverse pathways, which provides the opportunity to reveal how melatonin signal participates in plant resistance to cold stress. As previously described, melatonin can not only directly react with ROS to eliminate excessive ROS in plants, but also crosstalk with other signal molecules such as Ca^{2+}, NO, H_2O_2, BR, IAA, ABA, and JA to transmit cold signals. In addition, melatonin can also induce the expression levels of cold-responsive genes. In summary, melatonin affects the antioxidant systems, photosynthesis, and carbon assimilation processes of plants by regulating the expression levels of cold-related genes and interactions with other signaling molecules to improve the cold resistance of plants.

References

Aghdam MS, Luo Z, Jannatizadeh A et al (2019) Employing exogenous melatonin applying confers chilling tolerance in tomato fruits by upregulating *ZAT2/6/12* giving rise to promoting endogenous polyamines, proline, and nitric oxide accumulation by triggering arginine pathway activity. Food Chem 275:549–556
Aghdam MS, Luo Z, Li L et al (2020) Melatonin treatment maintains nutraceutical properties of pomegranate fruits during cold storage. Food Chem 303:125385
Ahn HR, Kim YJ, Lim YJ et al (2021) Key genes in the melatonin biosynthesis pathway with circadian rhythm are associated with various abiotic stresses. Plants (Basel) 10:129
Altaf MA, Shu HY, Hao YY et al (2022) Melatonin affects the photosynthetic performance of pepper (*Capsicum annuum* L.) seedlings under cold stress. Antioxidants 11:2414
Amasino RM (2005) Vernalization and flowering time. Curr Opin Biotechnol 16:154–158
Arnao MB, Hernandez-Ruiz J (2013) Growth conditions determine different melatonin levels in *Lupinus albus* L. J Pineal Res 55:149–155
Arnao MB, Hernandez-Ruiz J (2015) Functions of melatonin in plants: a review. J Pineal Res 59: 133–150
Arnao MB, Hernndez-Ruiz J (2014) Melatonin: plant growth regulator and/or biostimulator during stress? Trends Plant Sci 1–9:789–797
Auge GA, Blair LK, Neville H et al (2017) Maternal vernalization and vernalization-pathway genes influence progeny seed germination. New Phytol 216:388–400
Back K, Tan DX, Reiter RJ (2016) Melatonin biosynthesis in plants: multiple pathways catalyze tryptophan to melatonin in the cytoplasm or chloroplasts. J Pineal Res 61:426–437
Bajwa VS, Shukla MR, Sherif SM et al (2014) Role of melatonin in alleviating cold stress in *Arabidopsis thaliana*. J Pineal Res 56:238–245
Cao S, Bian K, Shi L et al (2018) Role of melatonin in cell-wall disassembly and chilling tolerance in cold-stored peach fruit. J Agric Food Chem 66:5663–5670
Chang J, Guo Y, Zhang Z et al (2020) CBF-responsive pathway and phytohormones are involved in melatonin-improved photosynthesis and redox homeostasis under aerial cold stress in watermelon. Acta Physiol Plant 42:159
Chang T, Zhao Y, He H et al (2021) Exogenous melatonin improves growth in *hulless barley* seedlings under cold stress by influencing the expression rhythms of circadian clock genes. PeerJ 9:e10740

Ding F, Liu B, Zhang S (2017) Exogenous melatonin ameliorates cold-induced damage in tomato plants. Sci Hortic 219:264–271

Ding F, Ren L, Xie F et al (2021) Jasmonate and melatonin act synergistically to potentiate cold tolerance in tomato plants. Front Plant Sci 12:763284

Fan J, Hu Z, Xie Y et al (2015) Alleviation of cold damage to photosystem II and metabolisms by melatonin in *Bermudagrass*. Front Plant Sci 6:925

Fan W, He Y, Guan X et al (2018) Involvement of the nitric oxide in melatonin-mediated protection against injury. Life Sci 200:142–147

Feng Y, Fu X, Han L et al (2021) Nitric oxide functions as a downstream signal for melatonin-induced cold tolerance in cucumber seedlings. Front Plant Sci 12:686545

Fu J, Zhang S, Jiang H et al (2022) Melatonin-induced cold and drought tolerance is regulated by brassinosteroids and hydrogen peroxide signaling in *perennial ryegrass*. Environ Exp Bot 196: 104815

Han QH, Huang B, Ding CB et al (2017) Effects of melatonin on anti-oxidative systems and photosystem II in cold-stressed rice seedlings. Front Plant Sci 8:785

Hardeland R (2015) Melatonin in plants and other phototrophs: advances and gaps concerning the diversity of functions. J Exp Bot 66:627–646

Hayat F, Sun Z, Ni Z et al (2022) Exogenous melatonin improves cold tolerance of strawberry (*Fragaria* × *ananassa* Duch.) through modulation of DREB/CBF-COR pathway and antioxidant defense system. Horticulturae 8:194

He H, He LF (2020) Crosstalk between melatonin and nitric oxide in plant development and stress responses. Physiol Plant 170:218–226

Hernandez IG, Gomez FJ, Cerutti S et al (2015) Melatonin in *Arabidopsis thaliana* acts as plant growth regulator at low concentrations and preserves seed viability at high concentrations. Plant Physiol Biochem 94:191–196

Hu Z, Fan J, Xie Y et al (2016) Comparative photosynthetic and metabolic analyses reveal mechanism of improved cold stress tolerance in bermudagrass by exogenous melatonin. Plant Physiol Biochem 100:94–104

Hu Z, Huang X, Amombo E et al (2020) The ethylene responsive factor CdERF1 from bermudagrass (*Cynodon dactylon*) positively regulates cold tolerance. Plant Sci 294:110432

Hwang OJ, Back K (2019) Melatonin deficiency confers tolerance to multiple abiotic stresses in rice via decreased brassinosteroid levels. Int J Mol Sci 20:5173

Janmohammadi M, Zolla L, Rinalducci S (2015) Low temperature tolerance in plants: changes at the protein level. Phytochemistry 117:76–89

Kang K, Lee K, Park S et al (2010) Enhanced production of melatonin by ectopic overexpression of human serotonin *N*-acetyltransferase plays a role in cold resistance in transgenic rice seedlings. J Pineal Res 49:176–182

Kong X, Ge W, Wei B et al (2020) Melatonin ameliorates chilling injury in green bell peppers during storage by regulating membrane lipid metabolism and antioxidant capacity. Postharvest Biol Technol 170:111315

Lei X-Y, Zhu R-Y, Zhang G-Y et al (2004) Attenuation of cold-induced apoptosis by exogenous melatonin in carrot suspension cells: the possible involvement of polyamines. J Pineal Res 36: 126–131

Lei Q, Wang L, Tan DX et al (2013) Identification of genes for melatonin synthetic enzymes in 'Red Fuji' apple (*Malus domestica Borkh. cv. Red*) and their expression and melatonin production during fruit development. J Pineal Res 55:443–451

Li X, Tan DX, Jiang D et al (2016) Melatonin enhances cold tolerance in drought-primed wild-type and abscisic acid-deficient mutant barley. J Pineal Res 61:328–339

Li H, Chang J, Zheng J et al (2017) Local melatonin application induces cold tolerance in distant organs of *Citrullus lanatus* L. via long distance transport. Sci Rep 7:40858

Li X, Brestic M, Tan D-X et al (2018) Melatonin alleviates low PSI-limited carbon assimilation under elevated CO_2 and enhances the cold tolerance of offspring in chlorophyll b-deficient mutant wheat. J Pineal Res 64:e12453

Li J, Yang Y, Sun K et al (2019) Exogenous melatonin enhances cold, salt and drought stress tolerance by improving antioxidant defense in tea plant (*Camellia sinensis* (L.) O. Kuntze). Molecules 24:1628

Li H, Guo Y, Lan Z et al (2021) Methyl jasmonate mediates melatonin-induced cold tolerance of grafted watermelon plants. Hortic Res 8:57

Liu C, Schlappi MR, Mao B et al (2019) The *bZIP73* transcription factor controls rice cold tolerance at the reproductive stage. Plant Biotechnol J 17:1834–1849

Liu G, Zhang Y, Yun Z et al (2020) Melatonin enhances cold tolerance by regulating energy and proline metabolism in litchi fruit. Foods 9:454

Ma C, Pei Z-Q, Bai X et al (2022) Exogenous melatonin and $CaCl_2$ alleviate cold-induced oxidative stress and photosynthetic inhibition in cucumber seedlings. J Plant Growth Regul 42(6):1–18

Manasa SL, Panigrahy M, Panigrahi KCS et al (2021) Overview of cold stress regulation in plants. Bot Rev 88:359–387

Miranda S, Vilches P, Suazo M et al (2020) Melatonin triggers metabolic and gene expression changes leading to improved quality traits of two sweet cherry cultivars during cold storage. Food Chem 319:126360

Mur LA, Mandon J, Persijn S et al (2013) Nitric oxide in plants: an assessment of the current state of knowledge. AoB Plants 5:pls052

Murch SJ, KrishnaRaj S, Saxena PK (2000) Tryptophan is a precursor for melatonin and serotonin biosynthesis in in vitro regenerated St. John's wort (*Hypericum perforatum* L. cv. Anthos) plants. Plant Cell Rep 19:698–704

Murch SJ, Alan AR, Cao J et al (2009) Melatonin and serotonin in flowers and fruits of *Datura metel* L. J Pineal Res 47:277–283

Paredes SD, Korkmaz A, Manchester LC et al (2009) Phytomelatonin: a review. J Exp Bot 60:57–69

Pu Y-J, Cisse EHM, Zhang L-J et al (2021) Coupling exogenous melatonin with Ca^{2+} alleviated chilling stress in *Dalbergia odorifera* T. Chen Trees 35:1541–1554

Qari SH, Hassan MU, Chattha MU et al (2022) Melatonin induced cold tolerance in plants: physiological and molecular responses. Front Plant Sci 13:843071

Raza A, Charagh S, Garcia-Caparros P et al (2022) Melatonin-mediated temperature stress tolerance in plants. GM Crops Food 13:196–217

Shafi A, Singh AK, Zahoor I (2021) Melatonin: role in abiotic stress resistance and tolerance. In: Aftab T, Hakeem KR (eds) Plant growth regulators: signalling under stress conditions. Springer, Cham, pp 239–273

Sharafi Y, Aghdam MS, Luo Z et al (2019) Melatonin treatment promotes endogenous melatonin accumulation and triggers GABA shunt pathway activity in tomato fruits during cold storage. Sci Hortic 254:222–227

Sharma A, Zheng B (2019) Melatonin mediated regulation of drought stress: physiological and molecular aspects. Plan Theory 8(7):190

Sharma A, Wang J, Xu D, Tao S, Chong S, Yan D et al (2020) Melatonin regulates the functional components of photosynthesis, antioxidant system, gene expression, and metabolic pathways to induce drought resistance in grafted Carya cathayensis plants. Sci Total Environ 713:136675

Shi H, Chan Z (2014) The cysteine2/histidine2-type transcription factor ZINC FINGER OF ARABIDOPSIS THALIANA 6-activated C-REPEAT-BINDING FACTOR pathway is essential for melatonin-mediated freezing stress resistance in *Arabidopsis*. J Pineal Res 57:185–191

Song L, Zhang W, Li Q et al (2022) Melatonin alleviates chilling injury and maintains postharvest quality by enhancing antioxidant capacity and inhibiting cell wall degradation in cold-stored eggplant fruit. Postharvest Biol Technol 194:112092

Sun H-J, Luo M-l, Zhou X et al (2020) Influence of melatonin treatment on peel browning of cold-stored "Nanguo" pears. Food Bioprocess Technol 13:1478–1490

Tan DX, Manchester LC, Esteban-Zubero E et al (2015) Melatonin as a potent and inducible endogenous antioxidant: synthesis and metabolism. Molecules 20:18886–18906

Turk H, Genisel M (2020) Melatonin-related mitochondrial respiration responses are associated with growth promotion and cold tolerance in plants. Cryobiology 92:76–85

Uchendu EE, Shukla MR, Reed BM et al (2013) Melatonin enhances the recovery of cryopreserved shoot tips of American elm (*Ulmus americana* L.). J Pineal Res 55:435–442

Wang M, Dai W, Du J et al (2019) ERF109 of trifoliate orange (*Poncirus trifoliata* (L.) Raf.) contributes to cold tolerance by directly regulating expression of Prx1 involved in antioxidative process. Plant Biotechnol J 17:1316–1332

Warne LGG (1947) Vernalization of lettuce. Nature 159:31

Wei LY, Liu CH, Wang JJ et al (2020) Melatonin immersion affects the quality of fresh-cut broccoli (*Brassica oleracea* L.) during cold storage: focus on the antioxidant system. J Food Process Preserv 44:e14691

Wu P, Ma Y, Ahammed GJ et al (2022) Insights into melatonin-induced photosynthetic electron transport under low-temperature stress in cucumber. Front Plant Sci 13:1029854

Xu R, Wang L, Li K et al (2022) Integrative transcriptomic and metabolomic alterations unravel the effect of melatonin on mitigating postharvest chilling injury upon plum (cv.Friar) fruit. Postharv Biol Technol 186:111819

Xu P, Huber DJ, Gong D et al (2023) Amelioration of chilling injury in 'Guifei' mango fruit by melatonin is associated with regulation of lipid metabolic enzymes and remodeling of lipidome. Postharvest Biol Technol 198:112233

Yang X, Ren J, Li J et al (2022) Meta-analysis of the effect of melatonin application on abiotic stress tolerance in plants. Plant Biotechnol Rep 17:39–52

Zhang YP, Yang SJ, Chen YY (2017) Effects of melatonin on photosynthetic performance and antioxidants in melon during cold and recovery. Biol Plant 61:571–578

Zhang Q, Li Y, Cao K et al (2023) Transcriptome and proteome depth analysis indicate ABA, MAPK cascade and Ca^{2+} signaling co-regulate cold tolerance in Rhododendron chrysanthum Pall. Front Plant Sci 14:1146663

Zhao C, Wang P, Si T et al (2017) MAP kinase cascades regulate the cold response by modulating ICE1 protein stability. Dev Cell 43:618–629

Zhou R, Wan HJ, Jiang FL et al (2020) The alleviation of photosynthetic damage in tomato under drought and cold stress by high CO_2 and melatonin. Int J Mol Sci 21:5587

Chapter 10
Physiological and Molecular Mechanisms of Melatonin-Induced Tolerance to Heavy Metal Stress

Yameng Wang, Keyin Shen, Shibei Ge, and Xin Li

Abstract Rapid development of the industry has increased the risk of environmental pollution, especially heavy metal pollution. When the crop uptakes large amounts of heavy metals from the soil, plant growth and development are severely hampered. Recently, the use of plant derivatives and other measures to alleviate heavy metal stress in plants has become a hot research topic. Almost all plants in nature contain melatonin, a small molecule indoleamine, which plays an important role in plant health because it promotes plant growth and helps plants better adapt to environmental stress. The exogenous application of melatonin plays a very important role in promoting the resistance of plants to environmental stress. Studies showed that the level of melatonin in plants increased under heavy metal stress, which was mainly due to the increased expression of genes related to melatonin synthesis in plants stimulated by heavy metals. Compared with only heavy metal stress, the plant height, biomass accumulation, and root activity increased with exogenous melatonin treatment under heavy metal stress, which confirmed that melatonin could significantly improve plant tolerance to heavy metal stress. Here, the main functions of melatonin in plants under heavy metal stress are reviewed, which can provide a reference for related research in the future.

Keywords Melatonin · Heavy metal · Antioxidant system · Plant growth · Resistance

---

Y. Wang · K. Shen · S. Ge · X. Li (✉)
Key Laboratory of Tea Quality and Safety Control, Ministry of Agriculture and Rural Affairs, Tea Research Institute, Chinese Academy of Agricultural Sciences, Hangzhou, People's Republic of China
e-mail: lixin@tricaas.com

A. Sharma, G. J. Ahammed (eds.), *Melatonin in Plants: Role in Plant Growth, Development, and Stress Response*, Plant Life and Environment Dynamics,
https://doi.org/10.1007/978-981-99-8051-2_10

10.1 Introduction

Metals with a density greater than 4.5 g/cm^3 are classified as heavy metals, including about 45 kinds, such as copper (Cu), lead (Pb), zinc (Zn), iron (Fe), mercury (Hg), cadmium (Cd), etc. At present, more and more wastewater, waste gas and slag are produced by industrial and mining enterprises, all of which contain heavy metals that have been entering the natural environment continuously. It is a matter of concern that the problem of heavy metal pollution is imminent (Ahammed and Yang 2022). Heavy metal pollution has attracted public attention because of its high enrichment, difficult degradation, and high toxicity. When plants are stressed by heavy metal, heavy metal ions are absorbed by plants, the enzyme activity of plant cells is inhibited, protein structures are destroyed, metabolic functions are disturbed, and growth and development are impaired, resulting in reduced yields, inhibited root elongation, and even death of plant in severe cases (Sharma et al. 2022).

Melatonin, also known as *N*-acetyl-5-methoxytryptamine, is a small molecule indoleamine. In 1958, melatonin was first extracted from the pineal gland of cattle by Lerner and Case (1959) and was shown to be a hormonal substance. An increasing number of studies have confirmed that melatonin is widespread in higher plants. Therefore, plant melatonin is gradually being recognized and studied by researchers. Up to now, substantial research has shown that melatonin is an important regulator of plant responses to abiotic and biotic stress. It has become a hot trend to apply melatonin to alleviate heavy metal stress in plants. In particular, the number of studies on melatonin-induced alleviation of heavy metal stress in plants is increasing rapidly (Fig. 10.1). With the development of research, knowledge about melatonin metabolic pathway in plants is becoming more and more known, and other substances in this pathway may also enhance the ability of plants to resist heavy metal stress. There is no denying that this is a new perspective (Yu et al. 2018). Treating pea plants with exogenous melatonin under the stress of heavy metal copper (Cu) improved the tolerance of plants to the heavy metal copper, and made the plants more self-healing (Tan et al. 2007). Research on how plant melatonin enhances the response of plants to heavy metal stress will not only help us to better understand the functions of melatonin in plant physiology and biochemistry, but also provide some ideas for the study of regulation mechanisms of metal detoxification in plants.

10.2 Phytomelatonin Content Under Heavy Metal Stress

For a long time, it was widely believed that melatonin, as a neuroendocrine hormone, was only widely found in animals. With further research, melatonin has been found to be a widely distributed small molecule active substance found in bacteria, microorganisms, fungi, algae, and most organisms in animals and plants (Tan et al. 2003). In plants, melatonin is mainly produced in chloroplasts and

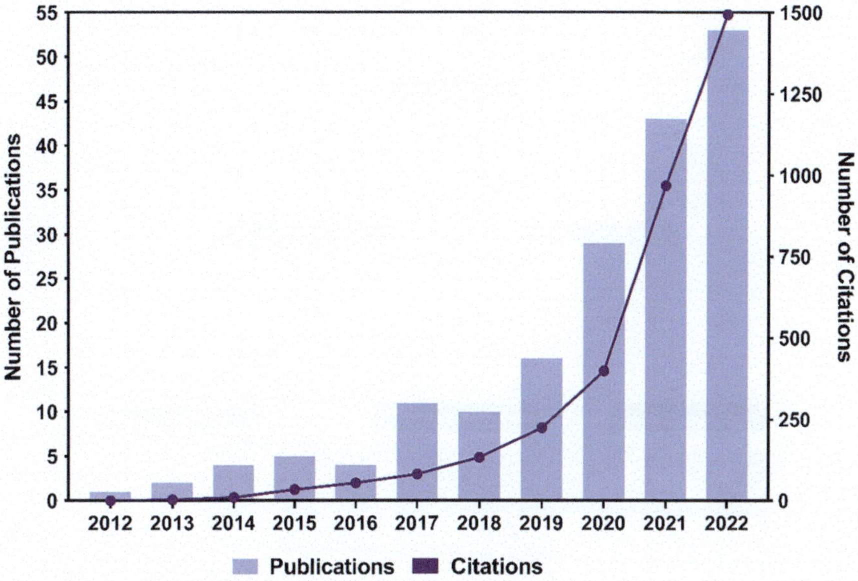

Fig. 10.1 The number of articles published on melatonin and heavy metal stress from the year 2012 to 2022 and relevant citations. Information was retrieved from the Web of Science with the keywords "melatonin" and "heavy metal stress"

mitochondria (Kovacik et al. 2017). Plants have essentially evolved complex cellular mechanisms to detoxify and tolerate heavy metals, such as immobilization, repulsion, chelation, segregation of metal ions, and repair of altered cell structure. When plants absorb heavy metals, the physiological metabolism of the plant undergoes a series of changes, and the levels of melatonin in the plant cell also respond to the rising metal levels (Fig. 10.2). The melatonin content of macroalgae was significantly increased under the stress of different heavy metals, such as Cd, Pb, Zn, etc. (Tal et al. 2011). This also demonstrates the mitigating effect of endogenous melatonin on heavy metal toxicity and the use of exogenous melatonin on plants to improve their ability to withstand heavy metal stress. Research has shown that treating barley and lupine (*Lupinus albus* L.) with zinc sulfate increases the endogenous melatonin content, which is related to the duration and concentration of metal treatment (Arnao and Hernandez-Ruiz 2009, 2013). Cd stress promoted melatonin synthesis in alfalfa (*Medicago sativa*) root tissues and attenuated Cd-induced growth inhibition in seedlings. Melatonin treatment promoted the re-establishment of redox homeostasis regulated by microRNAs and improved Cd tolerance in alfalfa. Research by Wu et al. (2019) found that there were a series of key genes for melatonin synthesis in plants. Under the stress of heavy metal Cd, the expression of these genes increased, and the content of endogenous melatonin in plants increased. Melatonin further activates the endogenous free radical scavenging system and specifically enhances the expression of two antioxidant enzymes, APX and SOD, thus improving the resistance to stress (Ni et al. 2018). In addition, Cd stress

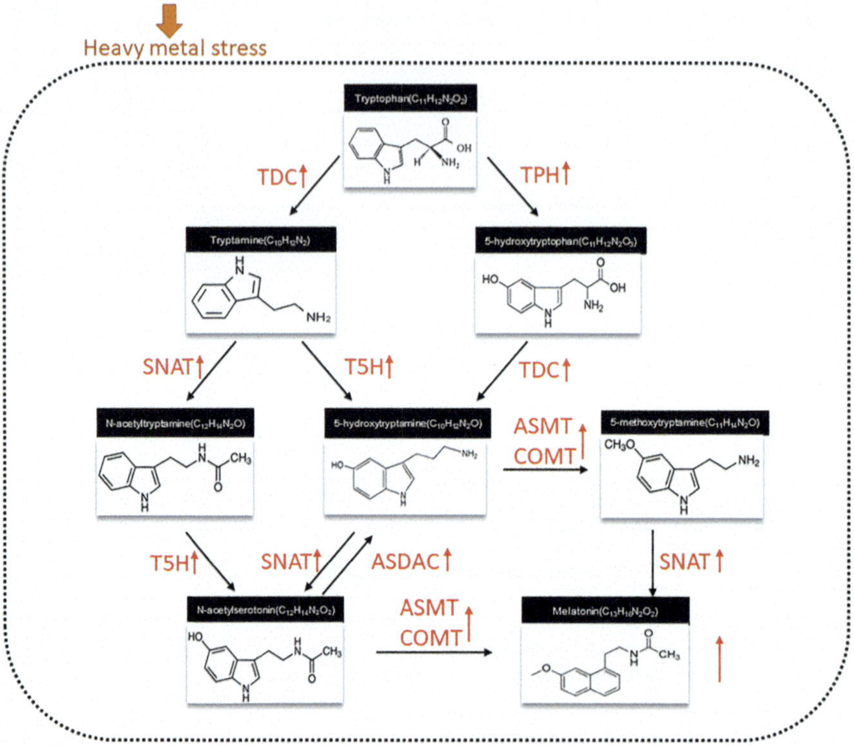

Fig. 10.2 Elevated melatonin content and enhanced expression of key synthetic genes within plants under heavy metal stress

induced the expression of *HSFA1a* and *HSP*s. The transcription factor protein Heat Shock Factor Ala (HSFAla) binds to the promoter of *COMT1*, a gene related to melatonin synthesis, and activates *COMT1* gene transcription to induce melatonin accumulation; partially upregulates the expression of *HSP*s, protects plasma membrane and intracellular proteins, and ultimately improves Cd tolerance in tomato (Cai et al. 2017). It was also found that melatonin significantly promoted the root growth of plant and enhanced the resistance of plants to adversity. Melatonin synthesis enzymes such as tryptophan decarboxylase (TDC), tryptamine 5-hydroxylase (T5H) and N-acetyl 5-hydroxytryptamine methyltransferase (ASMT) were significantly enhanced in rice under Cd stress, ultimately leading to increased melatonin synthesis and resistance to Cd stress (Byeon et al. 2015). Cd stress induced chloroplast damage can contribute to the massive synthesis of melatonin in rice leaves (Lee et al. 2018). Exogenous melatonin enhanced the synthesis of endogenous melatonin in watermelon and also induced the expression of genes such as superoxide dismutase, peroxidase, catalase, and glutathione transferase, reducing the accumulation of hydrogen peroxide and malondialdehyde. In addition, it reduced the transport of vanadium from roots to stems and leaves, thus decreasing the accumulation of vanadium in stems and leaves (Nawaz et al. 2018). In a word, when facing heavy metal stress, plants fight against stress by increasing internal metabolism. At this time, the expression of key genes in the hormone synthesis pathway such as melatonin synthesis pathway in the plant is increased and the endogenous melatonin content is enhanced, which is more effective in alleviating heavy metal stress in plants when melatonin is also administered exogenously. Not only heavy metal stress, but also other stresses increased endogenous melatonin in plants (Tettamanti et al. 2000), which also precisely confirmed that it is a new research idea to apply melatonin to relieve plants when they are under stress.

10.3 Mechanism of Melatonin Action in Alleviating Heavy Metal Stress in Plants

From a plant perspective, when plants are stressed by heavy metals, stress can be relieved in many ways and the specific mechanism is as follows.

10.3.1 Melatonin-Induced Heavy Metal Chelation and Vacuolar Sequestration

The first defense mechanism of the plant exposed to heavy metal stress is that PCs in the plant cytoplasm chelate heavy metals, and PC-metal complexes are segregated from plant vesicles (Clemens 2006; Limson et al. 1998). Plant chelating peptide is a kind of chelating peptide catalyzed by PC synthase with GSH as substrate, which is a natural ligand for heavy metal ions (de Oliveira et al. 2014). When the chelating peptide chelates with heavy metals, it can be transferred to the vacuole, reducing the toxicity of heavy metal ions (de Oliveira et al. 2016). Melatonin treatment promoted the biosynthesis of glutathione and phytochelatins and the vacuolar separation of cadmium under cadmium stress (Cai et al. 2017). Tal's study demonstrated that the addition of exogenous melatonin alleviated the toxic effects of Cd^{2+} stress on algae due to the ability of both melatonin and other precursors in the synthetic pathway to chelate heavy metals and also to induce the production of chelating peptides in plants. Therefore, the toxic effect of Cd on algae was reduced (Tal et al. 2011). Melatonin promotes the formation of tomato fruit cell walls under heavy metal stress (Sun et al. 2016). After exogenous melatonin treatment, the content of glutathione and plant chelating peptide in tomato plants increased significantly, and the proportion of Cd in cell walls and vacuoles in tomato roots and leaves increased under Cd^{2+} stress, which indicated that melatonin affected the migration and transportation of heavy metal ions in plants, thus alleviating the damage caused by Cd stress (Hasan et al. 2015). It was found that foliar spraying melatonin significantly improved the tolerance to Cd^{2+} tolerance in tobacco, which was due to the downregulation of Cd transporter by melatonin gene expression and the enhancement of Cd fixation in cell walls and vacuoles (Wang et al. 2019).

10.3.2 Promotion of Photosynthesis and Protection of Plant Photosystems

The ability of melatonin to maintain photosynthesis in plants under adverse conditions is also one of the main reasons for melatonin to promote plant growth. Melatonin and salicylic acid (SA) cooperate to reduce Cd absorption, increase photosynthetic pigment content, accelerate the AsA-GSH cycle, and regulate the glyoxalase system, thus enhancing the tolerance of plants to Cd stress (Amjadi et al. 2021). The research results of (*Catharanthus roseus* (L.) G. Don) showed that under Cd^{2+} stress, spraying melatonin on leaves significantly reduced electrolyte leakage, increased chlorophyll content and improved the dynamic balance of essential cations in *Catharanthus roseus* leaves (Nabaei and Amooaghaie 2019). Chloroplast damage induced by Cd stress contributes to the synthesis of melatonin in rice leaves (Lee and Back 2017). In regulating photosynthetic enzymes, it was found that La^{3+} stress inhibited photosynthesis and growth of tomato seedlings, and exogenous melatonin could maintain photosynthesis in tomato leaves by increasing leaf Rubisco activity and inhibiting reactive oxygen species formation (Siddiqui et al. 2019). Exogenous melatonin increased Pn, Gs and Tr of cucumber seedlings under Cd^{2+} stress, and chlorophyll fluorescence parameters indicated that the addition of melatonin alleviated the decrease in φPSII and Fv/Fm caused by Cd^{2+} stress (Wu et al. 2019). The root vigor and above-ground dry weight of Perilla (*Perilla frutescens* (L.) Britt.) increased after melatonin treatment under Cd^{2+} stress. At the same time, the content of chlorophyll a, chlorophyll b and total chlorophyll increased significantly (Xiang et al. 2019). In addition, melatonin increased chlorophyll content at the seedling stage and SPAD value at the flowering stage of wheat under Cr^{6+} stress. This may be related to the fact that melatonin downregulates chlorophyllase (*CLH1*) gene expression, which is related to chlorophyll degradation, downregulates senescence-related proteins, protects cellular ultrastructure, and maintains enzyme activities related to chlorophyll and photosynthesis (Byeon et al. 2013; Debnath et al. 2018; Weeda et al. 2014; Wei et al. 2015).

10.3.3 Melatonin Strengthens the Antioxidant System in Plants Under Stress

Melatonin is a strong antioxidant involved in many cellular activities in the plant body, effectively scavenging active oxygen and reducing MDA levels (Zhang et al. 2015). Melatonin is considered to directly scavenge reactive oxygen species and play an important role in protecting plants from oxidative damage (Poeggeler et al. 1996). In addition, melatonin has free radical scavenging properties, i.e., the products of melatonin-radical reaction, cyclic-3-hydroxymelatonin (c30HM), N-acetyl-N-formyl-5-methoxykynurenine (AFMK), and N-acetyl-5-methoxykynurenine (AMK), also have free radical scavenging functions (Reiter et al. 2014). The process of ROS scavenging by melatonin provides a large number of electrons. After losing electrons, melatonin is converted to AFMK, which has a stronger antioxidant effect than melatonin, and the synergistic effect of melatonin and AFMK enhances the scavenging effect of melatonin on ROS (Lee et al. 2016; Reiter et al. 2016). Exogenous melatonin treatment can enhance the activity of antioxidant enzymes and increase the content of antioxidant substances under heavy metal stress, decrease oxidation and increase plant biomass (stem length, root length and dry and fresh weight) (Sami et al. 2020). Studies of alfalfa revealed that melatonin pretreatment reduced the level of lipid peroxidation in the plants, and the contents of MDA and H_2O_2 also decreased. Melatonin maintained the dynamic balance of oxidation in alfalfa by upregulating the activities of antioxidant enzymes (such as SOD, GR, CAT and APX) (Antoniou et al. 2017). Melatonin can also be used as a signaling molecule, combined with upstream regulators, to improve the transcriptional levels of related genes, inducing the synthesis and activity of antioxidant enzymes, and thus improving the antioxidant capacity of the plant. Using melatonin to improve the ability of soybeans to resist Al^{3+} toxicity has been found to be promising, associated with melatonin enhancing the activity of antioxidant enzymes (Zhang et al. 2017). In studies on rice, after exogenous melatonin was added, the biomass of seedlings increased under Cd stress. Melatonin effectively reduced the inhibitory effects of Cd^{2+} stress on SOD and POD, decreased the MDA content, and alleviated the damage caused by Cd^{2+} stress (Lv et al. 2019).

10.3.4 Melatonin Promotes Plant Growth Under Heavy Metal Stress

When plants are poisoned by heavy metals, their growth is affected, and the plants become short and even die in serious cases. This poisoning can be alleviated by the exogenous application of melatonin. Melatonin promoted the growth of perilla under Cd^{2+} stress and significantly increased the root and above-ground biomass of perilla. Huang and colleagues sprayed radish under Cd^{2+} stress with different concentrations of melatonin, and found that spraying melatonin significantly increased the biomass of radish, among which 150 μmol L^{-1} melatonin had the best alleviation effect. Compared with the control, the biomass of roots, tubers, leaves, and edible parts of radish increased by 38.3%, 57.57%, 15.45%, and 29.99%, respectively (Huang et al. 2017). Studies on cucumber showed that exogenous melatonin reduced Cu toxicity, and adding exogenous melatonin could promote the growth of cucumber seedlings. Compared with Cu stress alone, the plant biomass, leaf area, root length, and root surface area increased significantly (Cao et al. 2019). Exogenous melatonin significantly increased the dry matter quality of roots, stems and leaves of safflower (*Carthamus tinctorius* L.) seedlings under Pb stress. This is because the exogenous addition of melatonin treatment significantly increased the content of plant PCs and reduced the uptake of Pb and root-to-crown translocation, thus reducing Pb toxicity (Namdjoyan et al. 2020). Melatonin significantly promoted the growth of rosemary (*Rosmarinus officinalis* L.) compared to As treatment alone. Exogenous addition of melatonin significantly reduced the As content in the above-ground parts and roots under As^{3+} stress, reducing the enrichment and transfer coefficients (Farouk and Al-Amri 2019).

10.3.5 Melatonin-Induced Regulation of Genes Associated with Stress Resistance

As shown in Figure 10.3, when plants suffer from stress, signal molecules such as reactive oxygen species, hormones, protein kinases, transcription factors, and related signal cascades play an important role in information transmission. Melatonin regulates the transcription levels of redox-related genes, conferring higher health indices and greater resistance to stress. The transcription level of redox-related genes is also affected by melatonin. It was found that melatonin significantly increased the gene expression levels of POD, GPX, and thioredoxin in rice and other plants (Byeon et al. 2013; Zhang et al. 2014). Melatonin also directly regulates specific resistance genes, such as *CLHl*, a chlorophyllase-related gene; *PAO*, a pheophorbide a oxygenase-related gene; and *NHX1/AKT1*, an ion homeostasis-related gene (Shi et al. 2015; Wang et al. 2013; Weeda et al. 2014). In addition, Cd stress induced the expression of *HSFA1a* and *HSPs* genes and promoted the biosynthesis of GSH and metal chelating peptides (PCs) (Cai et al. 2017). Exogenous melatonin can reduce the plant damage caused by Cd stress by upregulating the expression of yellow stripe 1-like (YSL), heavy metal ATPases (HMA) and ATP-binding cassette (ABC) transporter genes (Xu et al. 2020) and downregulating the expression levels of related genes *HMA3*, *HMA4* and *IRT2* (iron-regulated transporter) (Lv et al. 2019). Meanwhile, transgenic Arabidopsis overexpressing *SNAT* enhanced plant tolerance to Cd stress by upregulating the transcript levels of *ABC*, *PCR2,* and *HMA4* (Gu et al. 2017). Interestingly, melatonin alleviation of Cd stress was also associated with NO, which acts as a downstream signaling molecule of melatonin to regulate the expression of *IRT1* and *IRT2*, reducing the uptake and accumulation of Cd in oilseed rape (*B. campestris*) seedlings, thus improving the tolerance of plants to Cd (Wang et al. 2021). Exogenous application of melatonin upregulates the transcript levels of anthocyanin synthase (*ANS*) and chalcone synthase (*CHS*) genes to promote anthocyanin synthesis and reduce the toxicity of arsenic (As) and chromium (Cr) in tea (*Camellia sinensis*) and tomato (*Solanum lycopersicum*) (Li et al. 2021; Sun et al. 2023).

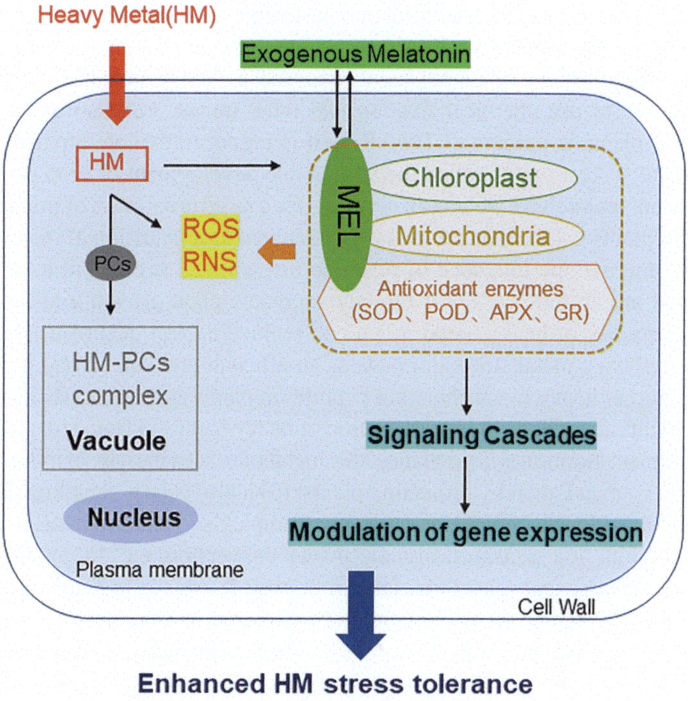

Fig. 10.3 Stress response of plants under heavy metal stress to exogenous melatonin treatment

10.4 Conclusions and Perspectives

With the detection of melatonin in plants for the first time, the study of melatonin has become more extensive and in-depth. Up to now, scholars at home and abroad have made outstanding achievements in the study of plant melatonin, which not only reveals and improves the synthetic route of plant melatonin, but also explores and studies various biological functions of plant melatonin. The research on melatonin has developed from the basic physiological and biochemical level to a more precise molecular and protein level. Studies have shown that melatonin can promote plant growth, which is closely related to the synergistic and antagonistic effects between melatonin and other plant hormones in the plant. Melatonin is also gradually used in horticultural plants, for example, as a plant growth regulator, preservative and anti-microbial agent to promote seed germination, regulate root structure, influence flowering and fruit set, prolong fruit preservation and enhance resistance to abiotic and biotic stresses in horticultural plants. In recent years, with the intensification of human activities, the rapid development of science and technology, the influence of mining and other factors, the phenomenon of heavy metal pollution in the world is becoming increasingly, which seriously threatens horticultural production. Heavy metals tend to accumulate in plants, and it is difficult to migrate in plants. Most importantly, it threatens the health of human beings and animals through the food chain, which is undoubtedly a serious potential threat. There is no reasonable solution to the problem of plants being exposed to heavy metal stress. It is not only expensive to use chemical fixation and other means, but also the impact on horticultural plants is unknown. Therefore, it is urgent to explore environmentally friendly plant protection to realize the sustainable development of crop production.

The current research on melatonin accounts for a large proportion of improving the tolerance of plants to heavy metal stress. The exogenous application of melatonin can effectively improve the tolerance of plants to heavy metal stress, and its mitigation mechanisms are as follows: (1) it directly promotes plant growth and helps them reduce the toxicity of heavy metal; (2) it can remove a great deal of active oxygen produced by heavy metal stress in plants as an efficient antioxidant; (3) it promotes the production of heavy metal chelating peptides in plants as a way to fix more heavy metals in plants and reduce the accumulation of heavy metals in fruits; (4) it synergies with other plant hormones to enhance the metabolic activities in plants under the stress of heavy metal, thereby protecting plants; (5) it also acts as signaling molecules directly involved in the connections between complex signaling networks in plants, such as inorganic ion transport, organic matter conversion, etc. In conclusion, the study of the features and roles of melatonin in plant resistance to heavy metal stress remains a priority and needs to continue to be explored in more depth.

The understanding of plant melatonin has been constantly updated along with research, but there are still the following difficulties in studying the specific role under heavy metal stress: (1) cultivating stable genetic horticultural plant varieties by using the existing molecular breeding methods is time-consuming and unpredictable; (2) precise control of melatonin dose under different heavy metal stress needs to be broken through; (3) functional verification of some key synthetic genes of melatonin is impossible; (4) the response pathways of plant endogenous melatonin under heavy metal stress are complicated, so it is difficult to accurately determine the receptors or related proteins of different signal responses. These problems have not yet been effectively solved. Not only in academic research, but also in the future, it is convenient to apply melatonin more effectively to plants tolerant to heavy metal stress and need more efforts to solve the existing problems in horticultural production. For example, how will plants coordinate melatonin levels in their bodies, so that they can cope with various types of stress most effectively? How to effectively improve melatonin levels in horticultural crops? Whether high stress tolerance in horticultural plants affects their quality or not? How to balance the relationship between melatonin levels in plants when they do affect the quality of horticultural plants? How to accurately control the amount of exogenous melatonin under heavy metal stress? These are worthy of more in-depth studies by scientists.

In summary, the research on plant melatonin in the future needs to continuously optimize innovative research tools. Combining with the current industry, an in-depth analysis of the toxicological mechanism of melatonin under heavy metal stress will help to improve the detoxification network of plants, provide theoretical basis and technical guidance for targeted breeding of horticultural plant varieties with high resistance and low accumulation, and carry out deep processing and product development.

Acknowledgments Research in the authors' laboratories was supported by the Innovation Project of the Chinese Academy of Agricultural Sciences (CAASASTIP-2019-TRICAAS).

References

Ahammed GJ, Yang Y (2022) Anthocyanin-mediated arsenic tolerance in plants. Environ Pollut 292:118475. https://doi.org/10.1016/j.envpol.2021.118475

Amjadi Z, Namdjoyan S, Abolhasani Soorki A (2021) Exogenous melatonin and salicylic acid alleviates cadmium toxicity in safflower (*Carthamus tinctorius* L.) seedlings. Ecotoxicology 30: 387–401

Antoniou C, Chatzimichail G, Xenofontos R, Pavlou JJ, Panagiotou E, Christou A, Fotopoulos V (2017) Melatonin systemically ameliorates drought stress-induced damage in Medicago sativa plants by modulating nitro-oxidative homeostasis and proline metabolism. J Pineal Res 62: e12401

Arnao MB, Hernandez-Ruiz J (2009) Chemical stress by different agents affects the melatonin content of barley roots. J Pineal Res 46:295–299

Arnao MB, Hernandez-Ruiz J (2013) Growth conditions determine different melatonin levels in Lupinus albus L. J Pineal Res 55:149–155

Byeon Y, Park S, Kim YS, Back K (2013) Microarray analysis of genes differentially expressed in melatonin-rich transgenic rice expressing a sheep serotonin *N*-acetyltransferase. J Pineal Res 55: 357–363

Byeon Y, Lee HY, Choi D-W, Back K (2015) Chloroplast-encoded serotonin N-acetyltransferase in the red alga Pyropia yezoensis: gene transition to the nucleus from chloroplasts. J Exp Bot 66: 709–717

Cai S-Y, Zhang Y, Xu Y-P, Qi Z-Y, Li M-Q, Ahammed GJ, Xia X-J, Shi K, Zhou Y-H, Reiter RJ, Yu J-Q, Zhou J (2017) HsfA1a upregulates melatonin biosynthesis to confer cadmium tolerance in tomato plants. J Pineal Res 62:e12387

Cao Y-Y, Qi C-D, Li S, Wang Z, Wang X, Wang J, Ren S, Li X, Zhang N, Guo Y-D (2019) Melatonin alleviates copper toxicity via improving copper sequestration and ROS scavenging in cucumber. Plant Cell Physiol 60:562–574

Clemens S (2006) Evolution and function of phytochelatin synthases. J Plant Physiol 163:319–332

de Oliveira LM, Ma LQ, Santos JAG, Guilherme LRG, Lessl JT (2014) Effects of arsenate, chromate, and sulfate on arsenic and chromium uptake and translocation by arsenic hyperaccumulator Pteris vittata L. Environ Pollut 184:187–192

de Oliveira LM, Gress J, De J, Rathinasabapathi B, Marchi G, Chen Y, Ma LQ (2016) Sulfate and chromate increased each other's uptake and translocation in As-hyperaccumulator Pteris vittata. Chemosphere 147:36–43

Debnath B, Hussain M, Irshad M, Mitra S, Li M, Liu S, Qiu D (2018) Exogenous melatonin mitigates acid rain stress to tomato plants through modulation of leaf ultrastructure, photosynthesis and antioxidant potential. Molecules 23:388

Farouk S, Al-Amri SM (2019) Exogenous melatonin-mediated modulation of arsenic tolerance with improved accretion of secondary metabolite production, activating antioxidant capacity and improved chloroplast ultrastructure in rosemary herb. Ecotoxicol Environ Saf 180:333–347

Gu Q, Chen Z, Yu X, Cui W, Pan J, Zhao G, Xu S, Wang R, Shen W (2017) Melatonin confers plant tolerance against cadmium stress via the decrease of cadmium accumulation and reestablishment of microRNA-mediated redox homeostasis. Plant Sci 261:28–37

Hasan MK, Ahammed GJ, Yin L, Shi K, Xia X, Zhou Y, Yu J, Zhou J (2015) Melatonin mitigates cadmium phytotoxicity through modulation of phytochelatins biosynthesis, vacuolar sequestration, and antioxidant potential in Solanum lycopersicum L. Front Plant Sci 6:601

Huang JJ, Lin LJ, Chen FB, Wei ZL, Wen K, Wang T, Liu L, Liao MA, Ren W (2017) Effects of Spraying Melatonin on Growth and Cadmium Accumulation of Radish Journal of Sichuan Agricultural University 35(3):375–380

Kovacik J, Rotkova G, Bujdos M, Babula P, Peterkova V, Matus P (2017) Ascorbic acid protects Coccomyxa subellipsoidea against metal toxicity through modulation of ROS/NO balance and metal uptake. J Hazard Mater 339:200–207

Lee K, Back K (2017) Overexpression of rice serotonin N-acetyltransferase 1 in transgenic rice plants confers resistance to cadmium and senescence and increases grain yield. J Pineal Res 62: e12392

Lee K, Zawadzka A, Czarnocki Z, Reiter RJ, Back K (2016) Molecular cloning of melatonin 3-hydroxylase and its production of cyclic 3-hydroxymelatonin in rice (Oryza sativa). J Pineal Res 61:470–478

Lee K, Lee HY, Back K (2018) Rice histone deacetylase 10 and Arabidopsis histone deacetylase 14 genes encode N-acetylserotonin deacetylase, which catalyzes conversion of N-acetylserotonin into serotonin, a reverse reaction for melatonin biosynthesis in plants. J Pineal Res 64:e12460

Lerner AB, Case JD (1959) Pigment cell regulatory factors. J Invest Dermatol 32:211–221

Li X, Ahammed GJ, Zhang X-N, Zhang L, Yan P, Zhang L-P, Fu J-Y, Han W-Y (2021) Melatonin-mediated regulation of anthocyanin biosynthesis and antioxidant defense confer tolerance to arsenic stress in Camellia sinensis L. J Hazard Mater 403:123922

Limson J, Nyokong T, Daya S (1998) The interaction of melatonin and its precursors with aluminium, cadmium, copper, iron, lead, and zinc: an adsorptive voltammetric study. J Pineal Res 24:15–21

Lv X, Fang Y, Zhang L, Zhang W, Xu L, Han J, Jin B, Zhang X, Zhang X, Xue D (2019) Effects of melatonin on growth, physiology and gene expression in rice seedlings under cadmium stress. Phyton Int J Exp Bot 88:91–100

Nabaei M, Amooaghaie R (2019) Nitric oxide is involved in the regulation of melatonin-induced antioxidant responses in Catharanthus roseus roots under cadmium stress. Botany 97:681–690

Namdjoyan S, Soorki AA, Elyasi N, Kazemi N, Simaei M (2020) Melatonin alleviates lead-induced oxidative damage in safflower (Carthamus tinctorius L.) seedlings. Ecotoxicology 29:108–118

Nawaz MA, Jiao Y, Chen C, Shireen F, Zheng Z, Imtiaz M, Bie Z, Huang Y (2018) Melatonin pretreatment improves vanadium stress tolerance of watermelon seedlings by reducing vanadium concentration in the leaves and regulating melatonin biosynthesis and antioxidant-related gene expression. J Plant Physiol 220:115–127

Ni J, Wang Q, Shah FA, Liu W, Wang D, Huang S, Fu S, Wu L (2018) Exogenous melatonin confers cadmium tolerance by counterbalancing the hydrogen peroxide homeostasis in wheat seedlings. Molecules 23:799

Poeggeler B, Reiter RJ, Hardeland R, Tan DX, BarlowWalden LR (1996) Melatonin and structurally-related, endogenous indoles act as potent electron donors and radical scavengers in vitro. Redox Rep 2:179–184

Reiter RJ, Tan DX, Galano A (2014) Melatonin: exceeding expectations. Physiology 29:325–333

Reiter RJ, Mayo JC, Tan D-X, Sainz RM, Alatorre-Jimenez M, Qin L (2016) Melatonin as an antioxidant: under promises but over delivers. J Pineal Res 61:253–278

Sami A, Shah FA, Abdullah M, Zhou X, Yan Y, Zhu Z, Zhou K (2020) Melatonin mitigates cadmium and aluminium toxicity through modulation of antioxidant potential inBrassica napusL. Plant Biol 22:679–690

Sharma A, Kapoor D, Gautam S, Landi M, Kandhol N, Araniti F et al (2022) Heavy metal induced regulation of plant biology: recent insights. Physiol Plant 174:e13688

Shi H, Jiang C, Ye T, Tan D-X, Reiter RJ, Zhang H, Liu R, Chan Z (2015) Comparative physiological, metabolomic, and transcriptomic analyses reveal mechanisms of improved abiotic stress resistance in bermudagrass [Cynodon dactylon (L). Pers.] by exogenous melatonin. J Exp Bot 66:681–694

Siddiqui MH, Alamri S, Alsubaie QD, Ali HM, Ibrahim AA, Alsadon A (2019) Potential roles of melatonin and sulfur in alleviation of lanthanum toxicity in tomato seedlings. Ecotoxicol Environ Saf 180:656–667

Sun Q, Zhang N, Wang J, Cao Y, Li X, Zhang H, Zhang L, Tan D-X, Guo Y-D (2016) A label-free differential proteomics analysis reveals the effect of melatonin on promoting fruit ripening and anthocyanin accumulation upon postharvest in tomato. J Pineal Res 61:138–153

Sun S, Liu A, Li Z, Guo T, Chen S, Ahammed GJ (2023) Anthocyanin synthesis is critical for melatonin-induced chromium stress tolerance in tomato. J Hazard Mater 453:131456. https://doi.org/10.1016/j.jhazmat.2023.131456

Tal O, Haim A, Harel O, Gerchman Y (2011) Melatonin as an antioxidant and its semi-lunar rhythm in green macroalga Ulva sp. J Exp Bot 62:1903–1910

Tan DX, Manchester LC, Hardeland R, Lopez-Burillo S, Mayo JC, Sainz RM, Reiter RJ (2003) Melatonin: a hormone, a tissue factor, an autocoid, a paracoid, and an antioxidant vitamin. J Pineal Res 34:75–78

Tan D-X, Manchester LC, Di Mascio P, Martinez GR, Prado FM, Reiter RJ (2007) Novel rhythms of N-1-acetyl-N-2-formyl-5-methoxykynuramine and its precursor melatonin in water hyacinth: importance for phytoremediation. FASEB J 21:1724–1729

Tettamanti C, Cerabolini B, Gerola P, Conti A (2000) Melatonin identification in medicinal plants. Acta Phytother 3:137–144

Wang P, Sun X, Chang C, Feng F, Liang D, Cheng L, Ma F (2013) Delay in leaf senescence of Malus hupehensis by long-term melatonin application is associated with its regulation of metabolic status and protein degradation. J Pineal Res 55:424–434

Wang M, Duan S, Zhou Z, Chen S, Wang D (2019) Foliar spraying of melatonin confers cadmium tolerance in Nicotiana tabacum L. Ecotoxicol Environ Saf 170:68–76

Wang T, Song J, Liu Z, Liu Z, Cui J (2021) Melatonin alleviates cadmium toxicity by reducing nitric oxide accumulation and IRT1 expression in Chinese cabbage seedlings. Environ Sci Pollut Res 28:15394–15405

Weeda S, Zhang N, Zhao X, Ndip G, Guo Y, Buck GA, Fu C, Ren S (2014) Arabidopsis transcriptome analysis reveals key roles of melatonin in plant defense systems. PLoS One 9: e93462

Wei W, Li Q-T, Chu Y-N, Reiter RJ, Yu X-M, Zhu D-H, Zhang W-K, Ma B, Lin Q, Zhang J-S, Chen S-Y (2015) Melatonin enhances plant growth and abiotic stress tolerance in soybean plants. J Exp Bot 66:695–707

Wu X, Li H, Luo H, Lin D, Zhou X (2019) Effect of exogenous melatonin on photosynthetic characteristics in cucumber seedlings under cadmium stress: a rapid detection method for the cadmium resistance. E3S Web Conf 136:07023

Xiang G, Lin L, Liao M, Tang Y, Liang D, Xia H, Wang J, Wang X, Sun G, Zhang H, Zou Y, Ren W (2019) Effects of melatonin on cadmium accumulation in the accumulator plant Perilla frutescens. Chem Ecol 35:553–562

Xu L, Zhang F, Tang M, Wang Y, Dong J, Ying J, Chen Y, Hu B, Li C, Liu L (2020) Melatonin confers cadmium tolerance by modulating critical heavy metal chelators and transporters in radish plants. J Pineal Res 69:e12659

Yu Y, Lv Y, Shi Y, Li T, Chen Y, Zhao D, Zhao Z (2018) The role of phyto-melatonin and related metabolites in response to stress. Molecules 23:1887

Zhang N, Zhang H-J, Zhao B, Sun Q-Q, Cao Y-Y, Li R, Wu X-X, Weeda S, Li L, Ren S, Reiter RJ, Guo Y-D (2014) The RNA-seq approach to discriminate gene expression profiles in response to melatonin on cucumber lateral root formation. J Pineal Res 56:39–50

Zhang N, Sun Q, Zhang H, Cao Y, Weeda S, Ren S, Guo Y-D (2015) Roles of melatonin in abiotic stress resistance in plants. J Exp Bot 66:647–656

Zhang J, Zeng B, Mao Y, Kong X, Wang X, Yang Y, Zhang J, Xu J, Rengel Z, Chen Q (2017) Melatonin alleviates aluminium toxicity through modulating antioxidative enzymes and enhancing organic acid anion exudation in soybean. Funct Plant Biol 44:961–968

Chapter 11
Melatonin Biology in Plants Under Biotic Stress: Physiological and Molecular Aspects

Tianmeng Guo, Huihang Xu, Fangying Zheng, Congrui Qin, and Shuangchen Chen

Abstract Melatonin is a multifunctional regulatory molecule and an important antioxidant. It is widely involved in various plant growth processes, such as circadian rhythm, growth and development, photosynthesis, anti-senescence, and so on. It is also known for enhancing the tolerance of plants to biotic stress. Melatonin can regulate the generation and reduction of reactive oxygen species, activate the plant antioxidant system, stimulate the accumulation of secondary metabolites, upregulate the expression of defense-related genes, and coordinate with other hormones (salicylates, jasmonates, ethylene, etc.) to confer resistance to biotic stressors of various kinds including fungi, bacteria, viruses, and insects. In this chapter, we review the functions of melatonin in plant immunity and the mechanism of melatonin-induced enhanced plant resistance to biotic stress from both physiological and molecular aspects.

Keywords Phytomelatonin · Plant immunity · Salicylic acid · Jasmonic acid · Secondary metabolism · Reactive oxygen species

T. Guo · H. Xu · F. Zheng · C. Qin
College of Horticulture and Plant Protection, Henan University of Science and Technology, Luoyang, People's Republic of China

S. Chen (✉)
College of Horticulture and Plant Protection, Henan University of Science and Technology, Luoyang, People's Republic of China

Henan International Joint Laboratory of Stress Resistance Regulation and Safe Production of Protected Vegetables, Luoyang, People's Republic of China

Henan Engineering Technology Research Center for Horticultural Crop Safety and Disease Control, Luoyang, People's Republic of China

A. Sharma, G. J. Ahammed (eds.), *Melatonin in Plants: Role in Plant Growth, Development, and Stress Response*, Plant Life and Environment Dynamics,
https://doi.org/10.1007/978-981-99-8051-2_11

11.1 Introduction

Melatonin (MT) is a small molecule, an indole derivative, essential for life, with the chemical name *N*-acetyl-5-methoxytryptamine (Li et al. 2022). It exists in the pineal gland of cattle and was first discovered in 1958 (Lerner et al. 1958). At first, researchers believed that melatonin existed only in animals, and it was not until 1995 that it was first detected in higher plants such as morning glory and tomato fruit (Vantassel et al. 1995). Since then, a boom in human research on melatonin has been set off. Natural melatonin is mainly divided into two types, namely melatonin from animals and melatonin from plants, the latter also known as "phytomelatonin" (Arnao and Hernández-Ruiz 2018). Melatonin exists in almost all plants and is distributed throughout various organs of plants. Its concentration varies among different species or different varieties or tissues of the same species, which is determined by its location, plant growth period, tissue, and harvesting time (Sharif et al. 2018b; Wu et al. 2021). However, the consensus is that its content is relatively low, usually pg/g-ng/g (Wu et al. 2021; Li et al. 2022).

Melatonin produced in plants mainly comes from chloroplasts and mitochondria (Wang et al. 2017; Zhao et al. 2019a), but currently, the chloroplast pathway is considered the most important pathway for melatonin synthesis (Ahammed and Li 2022; Zeng et al. 2022). After the synthesis of melatonin, it is transported to other locations in plants, thereby exerting corresponding effects (Wang et al. 2016). The basic steps of biosynthesis of melatonin have been discovered so far. Researchers used isotope labeling to supply tryptophan to various plants in vitro (Murch et al. 2000). The biosynthetic pathways of melatonin in animals and plants are similar, and the precursor substance is tryptophan. Its synthetic pathway is tryptophan converted by four consecutive enzyme catalysis. Plants use tryptophan as a substrate to generate tryptamine under the action of tryptophan decarboxylase (TDC), tryptamine is converted into *N*-acetyltryptamine through acetylation of serotonin *N*-acetyltransferase (SNAT), and then *N*-acetyl-5-hydroxytryptamine is synthesized through hydroxylation of tryptamine-5-hydroxylase (T5H). Subsequently, *N*-acetyl-5-hydroxytryptamine is catalyzed by *N*-acetylserotonin *O*-methyltransferase (ASMT) or caffeic acid *O*-methyltransferase (COMT) to produce melatonin (Zhao et al. 2019a; Zeng et al. 2022; Hernández-Ruiz et al. 2023). Among these enzymes, TDC, ASMT, and COMT are key rate limiting enzymes in the plant melatonin synthesis pathway.

More and more studies have shown that melatonin not only affects normal physiological functions, such as circadian rhythm, growth and development, photosynthesis, and anti-aging, etc. (Sharif et al. 2018b; Nawaz et al. 2020), but also plays a crucial role in improving the ability of plants to resist biotic stress. Here, we summarize the main functions of melatonin in plant defense and the mechanism of enhancing plant tolerance to biotic stress from both physiological and molecular aspects.

11.2 Melatonin Promotes Plant Tolerance to Biotic Stress

Biotic stress refers to the adverse effects of fungi, bacteria, viruses, and other organisms on plants. If a plant is infected with fungi or bacteria, it may exhibit leaf spots, stem vein decay, and endanger various organs of the plant, restricting the quality and yield of the plant. If infected with the virus, it can cause local plant lesions and systemic damage, ultimately leading to leaf deformities, developmental delays, and chlorosis (Moustafa-Farag et al. 2019). In a word, biotic stress can seriously limit crop production and cause huge losses. In the process of resisting biotic stress, melatonin plays a crucial role in immune regulation. Melatonin is an environmentally friendly molecule and is considered to be one of the most economical and green regulators to induce plant resistance to biotic stress (Moustafa-Farag et al. 2019).

Melatonin has a certain inhibitory effect on fungal diseases. Melatonin root irrigation treatment can significantly improve the immune ability of apple trees to fungi (*Diplocarpon mali*), specifically manifested by increased photosynthesis (Yin et al. 2013). After overexpression of the melatonin synthesis gene *SNAT1*, transgenic watermelon plants rich in melatonin showed stronger tolerance to powdery mildew (Mandal et al. 2018). Melatonin has broad-spectrum antifungal activity and can alleviate the damage caused by fungal infections on *Solanum lycopersicum* (Liu et al. 2019), *Cucumis sativus* (Ahammed et al. 2020), *Nicotiana benthamiana* (Zhang et al. 2018), etc. Surprisingly, melatonin and fungicides can also form synergistic effects, increase the sensitivity of fungicides, reduce the resistance of *Phytophthora infestans*, improve the prevention and control effect of fungicides, and reduce the use of fungicides (Zhang et al. 2017). In addition, in vitro experiments have found that melatonin pretreatment can significantly reduce the fungal infection rate during seed germination and inhibit the growth of fungal pathogens (Arnao and Hernández-Ruiz 2015). Overall, melatonin alleviates the damage caused by fungal infections by regulating plant resistance, reducing pathogen invasion, and hindering pathogen spread.

Melatonin enhances plant resistance to bacteria. At present, the most widely studied model for plant-bacterial interaction is *Arabidopsis thaliana/Pseudomonas syringae* pv. tomato DC3000 (Pst DC3000) (Arnao and Hernández-Ruiz 2017). In *Arabidopsis thaliana*, the ability of *SNAT* gene knockout mutants to resist lilac is reduced due to reduced expression of defense genes, and exogenous melatonin can effectively enhance the expression of defense genes (Lee et al. 2015). After melatonin treatment, the content of callose, cellulose, and some sugars in plants increases, causing cell wall thickening and thereby inducing enhanced plant resistance. From this, it can be seen that melatonin enhances the plant's resistance to bacteria. The bactericidal ability of melatonin against Gram multidrug-resistant bacteria, such as *Pseudomonas aeruginosa*, *Staphylococcus aureus*, and so on, has also been fully demonstrated (Chen et al. 2018). Similarly, melatonin enhances cassava's resistance to bacterial wilt by inducing transcription factors (*MeRAV2* and *MeRAV1*) (Wei et al. 2018b).

Compared to fungi and bacteria, if plants are infected with viruses, it is difficult to control and thus pose a huge threat to plants. Melatonin has been found to be antiviral in animals. In plants, it was also found that root irrigation with melatonin could effectively reduce the tobacco mosaic virus (TMV) titer in tobacco and tomato seedlings. The reason for this positive effect is the increase of nitric oxide (NO)-dependent salicylic acid (SA) synthesis in the defense pathway (Zhao et al. 2019b). In addition, melatonin can control apple stem groove disease caused by apple stem groove virus (ASGV) (Chen et al. 2019a) and reduce the incidence rate of rice virus disease (Lu et al. 2019). As research on the interaction between plants and viruses continues, melatonin may point out new directions for the prevention, treatment, and control of plant viral infections. Table 11.1 illustrates the effect of exogenous melatonin treatment on plant biotic stress tolerance.

Insects are the most notorious source of biological stress in the global crop production system. Its harm mainly lies in laying eggs or eating. For example, during the feeding process, stinging insects transmit viruses to plants through their probes, thereby causing harm to plants (Moustafa-Farag et al. 2019). At present, research shows that melatonin can improve the tolerance of plants to insects by influencing the seasonal rhythm, photoperiod, molting, reproduction, and circadian rhythm oscillation of some insects (Tiwari et al. 2021). A study on the pea aphid suggests that melatonin may have an impact on its photoperiodic response. Exogenous melatonin can induce the expression of its free radical defense genes, thereby improving the ability of citrus to resist Huanglong disease (Nehela and Killiny 2018). At present, dopamine has a positive impact on plant defense against herbivores, and the structure of melatonin is similar to dopamine, so it is speculated that the two have similar functions.

During the long-term coexistence of plants and pathogenic microorganisms, plants form two unique immune systems to resist the invasion of pathogenic microorganisms (Song et al. 2021). The first layer of the immune defense system relies on pattern recognition receptors (PRRs) on the cell membrane, which trigger the innate immune response by recognizing some foreign microbial associated molecular pattern (MAMP) or endogenous damage associated molecular pattern (DAMP), which leads to basal immunity called pattern recognition receptor triggered immunity (PTI)(Moustafa-Farag et al. 2019). However, some pathogens can secrete effector proteins to defeat PTI, at which point the second line of defense is activated. At this point, in order to overcome the invasion of pathogenic microorganisms on plants, plants have evolved nucleotide-binding (NB) leucine rich repeat (LRR) domain receptors (NLRs), which can specifically recognize and bind effector proteins to activate effector-triggered immunity (ETI) triggered by the second layer of effector proteins (Yuan et al. 2021). Compared to PTI, ETI mediates a more intense immune response, typically accompanied by programmed cell death (PCD), the so-called the hypersensitive response (HR) (Moustafa-Farag et al. 2019). Hypersensitivity reactions can cause rapid and massive accumulation of NO and reactive oxygen species (ROS), which can starve pathogenic microorganisms to death in local spaces and prevent further infection of adjacent cells. This can also be considered a marker reaction for ETI.

Table 11.1 Effects of exogenous melatonin treatments on biotic stress tolerance in plants

Plant species	Pathogen names	Type of pathogens	Disease name and type	Concentrations and methods of melatonin treatment	Effect on disease resistance	References
Potato (*Solanum tuberosum* L.)	*Phytophthora infestans*	Fungus	Potato late blight (foliar disease)	Seed treatments (6.0 mM, 10.0 mM)	Increased resistance to *P. infestans*	Zhang et al. (2017)
					Decreased potato late blight	
Radish (*Raphanus sativus L.*)	*Alternaria brassicae*	Fungus	Radish black spot C (foliar disease)	Foliar treatment (500 μM)	Increased resistance to *A. brassicae* significantly less blight	Zhang et al. (2023)
Malus prunifolia	*Diplocarpon mali*	Fungus	Apple brown spot (foliar disease)	Root irrigation (50–500 μM)	Reduced disease damage; increased resistance, and reduced pathological changes	Yin et al. (2013)
Musa acuminata	*Fusarium oxysporum*	Fungus	Banana fusarium wilt (fruit disease)	Leaf and root treatment (100 μM)	Improvement of disease resistance	Wei et al. (2017)
Fragaria ananassa	*Botrytis cinerea and rhizopus stolonifer*	Fungus	Strawberry gray mold (fruit disease)	Fruit dipping (100 μM)	Reduced the degree of fruit decay after picking	Aghdam and Fard (2017)
Citrullus lanatus	*Podosphaera xanthii and phythophthora capsici*	Fungus	Watermelon powdery mildew (foliar disease)	Leaf treatment (100 μM)	Enhanced disease resistance	Mandal et al. (2018)
Nicotiana glutinosa and solanum lycopersicum	*Tobacco mosaic virus* (TMV)	Virus	Tobacco virus disease; tomato mosaic virus disease (foliar disease)	Root irrigation 100 μM (twice)	Reduction of virus concentration in infected plants	Zhao et al. (2019b)
Malus domestica	*Apple stem grooving virus* (ASGV)	Virus	Apple stem furrow disease (foliar disease)	To the shoot proliferation medium (15 μM)	Eradication of virus from previously infected shoot tips	Chen et al. (2019a)

(continued)

Table 11.1 (continued)

Plant species	Pathogen names	Type of pathogens	Disease name and type	Concentrations and methods of melatonin treatment	Effect on disease resistance	References
Eggplant (*Solanum melongena*)	*Alfalfa mosaic virus* (AMV)	Virus	Eggplant virus disease (foliar disease)	Foliar spray (100 μM)	Maintained oxidation-reduction homeostasis and reduced reactive oxygen species	Sofy et al. (2021)
A. thaliana and *N. benthamiana*	*Pseudomonas syringae* pv. *tomato* DC3000	Bacteria	Angular leaf spot of tobacco (foliar disease)	Leaf treatment (1 μM)	Enhanced disease resistance	Lee and Back (2016)
Cassava (*Manihot esculenta Crantz*)	*Xanthomonas axonopodis* pv. *Manihotis*	Bacteria	Cassava bacterial blight (foliar disease)	Leaf treatment (50, 100, 150 μM)	Increased disease resistance by ethylene	Wei et al. (2022)
Oryza sativa and *N. benthamiana*	*Xanthomonas oryzae* pv. *oryzae*	Bacteria	Rice bacterial blight (foliar disease)	Foliar spray (861 μM)	Reduced pathogenicity and incidence of disease	Chen et al. (2019b)

In the process of biotic stress, when PRRs recognize PAMPs/MAMPs, PTI is triggered. At the same time, some effectors secreted by pathogens trigger ETI after being recognized by the host's R gene. At this time, ROS and RNS are produced, triggering the increase of melatonin content in plants. Melatonin directly or indirectly promote the expression of various intracellular and extracellular resistance proteins required for ETI and PTI-mediated resistance signals. Meanwhile, it induces the expression of various plant defense immune genes, activate defense-related enzymes and induce antibacterial compounds, to enhance the resistance of plants to biotic stress and protect plants from serious attacks by pathogenic bacteria. In gourds, melatonin is associated with the upregulation of host powdery mildew defense-related genes mediated by PTI and ETI. In watermelon leaves sprayed with melatonin (1 mM), multiple defense genes involved in PAMP and ETI were significantly induced (Mandal et al. 2018).

11.3 Reactive Oxygen Species (ROS), Reactive Nitrogen Species (RNS), and Antioxidative Defense in Melatonin-Induced Biotic Stress Tolerance

ROS and RNS can usually maintain plant redox homeostasis at lower levels. But in the early stages of biotic stress, redox homeostasis is often disrupted, leading to excessive production of ROS (singlet oxygen: $^1O_2$, superoxide anion: $O_2^{\cdot-}$, and hydrogen peroxide: H_2O_2) and RNS (nitric oxide: NO and alkoxy: RO), known as "oxidative burst," which is a major and widespread plant defense and stress signal response (Tiwari et al. 2021). Under abiotic stress, if there is an oxidative burst, the degree of oxidation in plants continues to decrease. When plants recognize the entry of pathogens, they trigger defense responses, activating local and systemic defenses, inducing local HR, and inducing local plant PCD at the infected site, ultimately preventing pathogens from continuing to enter the plant cells (Zhao et al. 2021). When infected with pathogenic bacteria, melatonin induces the accumulation of ROS and RNS in plants, promoting cell death and preventing further invasion. At the same time, the accumulated ROS and RNS also stimulate the production of melatonin, and there is also a certain interaction between ROS and RNS. Therefore, any one of the substances in melatonin, ROS, and RNS can promote the accumulation of the other two substances, thus combining into a positive regulatory cycle (Arnao and Hernández-Ruiz 2019; Zhao et al. 2021). This cycle works upstream of the plant's immune system and can protect the plant by activating and regulating the plant's HR, regulating biological defense signals located downstream (e.g., the MAPK cascade, plant hormone pathways, etc.).

In the late stage of biotic stress, when pathogen infection is suppressed, the excess ROS and RNS produced, if not cleared in a timely manner, can cause irreversible damage to cells. Meanwhile, it is necessary to activate antioxidant defense mechanisms to neutralize the excess ROS and RNS. Melatonin is a strong antioxidant, and its metabolites are directly related to ROS and RNS, which can directly and efficiently eliminate excess ROS and RNS. In the meantime, it can also induce non-enzymatic antioxidants such as ascorbic acid (AsA), glutathione (GSH), carotenoids, and phenolic compounds and activate antioxidant enzymes to function, such as superoxide dismutase (SOD), guaiacol peroxidase (POD), glutathione-*S*-transferase (GST), and ascorbate peroxidase (APX) antioxidant systems to indirectly eliminate excess ROS and RNS (Moustafa-Farag et al. 2019). Melatonin has a strong antioxidant capacity, twice that of vitamins, four times that of glutathione, and even 14 times that of mannitol. A single molecule of melatonin can clear up to ten times the amount of ROS or RNS.

Under biotic stress, melatonin can increase the content of NO by upregulating nitrate reductase (NR), thus improving plant stress resistance. For example, in the fight against TMV, melatonin can promote the generation of NO and SA to enhance the local resistance of tobacco and the systemic resistance of plants. Research shows that melatonin can promote the synthesis of sugars and glycerol in plants, increase endogenous NO levels, and thus enhance the immunity of *Arabidopsis thaliana* to bacterial pathogens (Qian et al. 2015). Generally, ROS and RNS are relatively stable and are transmitted over long distances to improve plant tolerance, while NO is transmitted over short distances. NO can self-regulate, while RNS and ROS mostly function in a coordinated manner. NO is also believed to be involved in modifying and regulating the ROS network, and the two are positively correlated, nonetheless, melatonin crosstalk with ROS and NO to fine-tune defense responses (Fig. 11.1).

11.4 Melatonin Stimulates Plant Secondary Metabolism to Enhance Biotic Stress Tolerance

Polyphenol oxidase (PPO) is an important enzyme that plays a crucial role in resisting the invasion of pathogenic microorganisms. It catalyzes the formation of lignin and quinone compounds, forming a protective barrier that protects cells from bacterial invasion. Phenylalanine ammonia-lyase (PAL) is an indispensable enzyme in the phenylpropane metabolism pathway, and the lignin involved in the formation can increase cell wall thickness and form a mechanical barrier for pathogen invasion (Aghdam and Fard 2017). The application of melatonin to apple fruits inoculated with *Botrytis cinerea* can significantly induce the activities of POD, CAT, SOD, PPO, and PAL in the apple fruits, thereby reducing the incidence of apple gray mold (Liu et al. 2019). Root application of melatonin in the rhizosphere can significantly improve the activity of POD, PAL, chitinase, and β-1,3-glucanase, thereby enhancing the resistance of apples to brown spot (Yin et al. 2013). Chitinase is an important substance that limits the continued growth and spread of pathogens, and melatonin can enhance its activity (Sharif et al. 2018a), and the same results have been verified in resistance against-anthracnose in peppers (Ali et al. 2020). Melatonin can inhibit

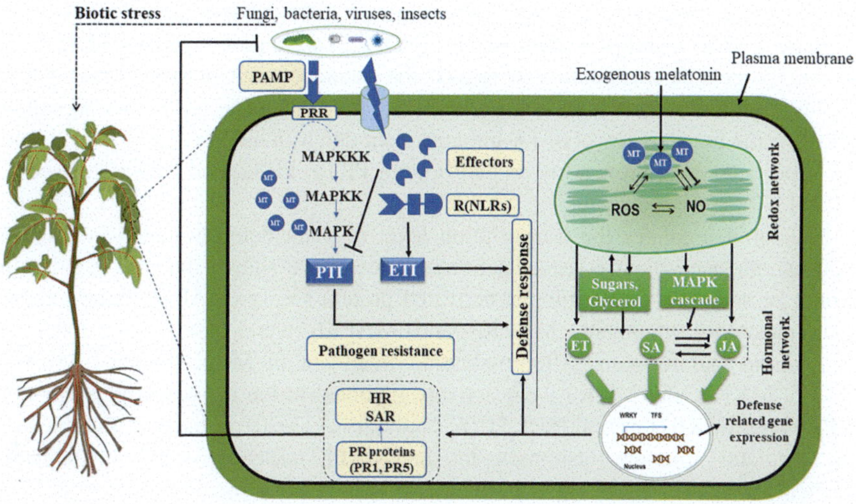

Fig. 11.1 Mechanism of melatonin-induced biotic stress tolerance. *MT* melatonin, *ROS* reactive oxygen species, *NO* nitric oxide, *ET* ethylene, *SA* salicylic acid, *JA* jasmonic acid, *TFs* transcription factors, *HR* hypersensitive response, *SAR* systemic acquired resistance, *PTI* pattern recognition receptor triggered immunity, *ETI* effector-triggered immunity, *MAPK* mitogen-activated protein kinase, *NLRs* nucleotide-binding leucine rich repeat domain receptors, *PAMP* pathogen-associated molecular pattern, *PR* pathogenesis-related protein, *PRR* pattern recognition receptors. (Modified and redrawn from Hernández-Ruiz et al. (2023). Created with BioRender.com)

the spread of infection spots caused by *Peronophytora lichii* in lychee fruits and induce resistance to pathogenic fungi in lychee by regulating the phenylpropanoid metabolism pathway, pentose phosphate pathway, and energy conversion (Zhang et al. 2021). A previous study has shown that melatonin enhances cotton's resistance to Fusarium wilt by inducing the upregulation of key genes in phenylalanine, valproate, and gossypol synthesis (Li et al. 2019).

Soluble proteins and soluble sugars are important substances that maintain plant cell balance. They can prevent cell membranes from being damaged by pathogens and control material exchange inside and outside cells, thereby reducing damage caused by pathogen infection. Melatonin treatment can significantly increase the soluble protein content of cucumber plants when infected with downy mildew fungi (Sun et al. 2019). Similarly, melatonin induces the transcription level of *CBF/DREB1s* in plants, promoting the synthesis of soluble sugars such as sucrose, thereby enhancing plant tolerance to bacterial pathogens (Shi et al. 2015).

As mentioned above, melatonin plays a specific role in the defense against the virulent pathogenic bacterium *P. syringae* DC3000. Melatonin can induce multiple disease resistance-related genes (PRs) and defense genes related to SA and ethylene in *Arabidopsis thaliana* and tobacco leaves. In *Arabidopsis thaliana SNAT* gene mutants, the expression of disease course defense-related genes was affected. When inoculated with *P. syringae* DC3000, the content of melatonin synthesis in the mutant of *SNAT* decreased to 50% of the wild type, and the sensitivity to pathogen infection was enhanced. Moreover, the expression of defense gene *PR1*, isomer synthase 1 (*ICS1*), and plant defensin 1.2 (*PDF1.2*) were downregulated (Lee et al. 2015). Exogenous melatonin can reduce the incidence rate of rice bacterial leaf blight by inducing the expression of *NPR1*, a regulatory gene on the SA signaling pathway, the transcript level of JA-induced plant defensin marker gene *PDF1.2*, and the expression of pathogenesis-related protein PR1b, PR8a, and PR9 (Chen et al. 2019b).

Melatonin regulates the transcription level of plant defense-related genes and induces enhanced plant resistance. Some studies have shown that this regulation cannot be separated from mitogen-activated protein kinase (MAPK). After exogenous melatonin treatment, MAPK3 and MAPK6 were rapidly expressed in *A. thaliana* and tobacco. After melatonin treatment of MKK mutants, MKK4, MKK5, MKK7, and MKK9 were actively involved in the activation pathways of MPK3 and MPK6 (Lee and Back 2016; Nawaz et al. 2020), indicating that melatonin regulates immune responses in the MAPK cascade signaling pathway (Hernández-Ruiz et al. 2023).

11.5 Melatonin Crosstalk with Other Hormones to Confer Biotic Stress Tolerance

Under biotic stress, plants often elevate levels of endogenous defense hormones such as SA, jasmonic acid (JA), and ethylene. Research shows that melatonin regulates plant resistance to biotic stress by interacting with signal transduction pathways regulated by SA, JA, NO, etc. The accumulation of SA activated by melatonin is crucial for the defense response of *Arabidopsis thaliana* against pathogen attacks triggered by melatonin, indicating a direct relationship between melatonin and SA in plant immunity. Melatonin upregulates the expression of SA and ethylene-related genes, but this effect is inhibited in SA and ethylene-signaling deficient mutants. The *SNAT* gene knockout not only reduces melatonin content, but also SA content, thereby reducing its resistance to pathogens (Lee et al. 2015). The enhanced resistance of apples to spot disease is also due to the significant increase in SA content in the apple after melatonin treatment (Yin et al. 2013).

Jasmonic acid is also essential for plants to resist biotic stress. According to reports, endogenous melatonin levels can regulate the JA signaling pathway to help tomatoes defend against gray mold (Liu et al. 2019). Melatonin induces the accumulation of JA by increasing the transcription level of PbOPR3, which enhances pear resistance to ring rot by stimulating ethylene and upregulation of PR and defense genes (Li et al. 2022). While ethylene enhances the infectivity of cucumber mosaic virus (CMV) in cucumbers and accelerates disease development, melatonin treatment can significantly inhibit ethylene activity and enhance plant defense against this specific virus (Sharif et al. 2018b). RNA-seq studies have shown that melatonin treatment on watermelon leaves induces significant upregulation of various plant hormone defense-related genes, including SA, JA, and ethylene (Mandal et al. 2018), with similar results in *Arabidopsis thaliana* (Weeda et al. 2014).

Biotic stress can induce plants to produce two kinds of resistance reactions, namely, the local resistance of plant-infected parts against pathogens and the systemic acquired resistance (SAR) of infected parts, which protects the uninjured plant tissues at the uninfected parts at the distal end. Local damage causes a plant immune response, producing defense signaling molecules that move to distant tissues to activate the plant's resistance to secondary infection. Therefore, local resistance is the basis for generating SAR. The specific microbial strain-induced plant resistance is known as induced systemic resistance (ISR). These two resistance responses are produced by multiple signaling factors regulating defense mechanisms in different pathways, and ethylene, JA, and SA are prominent in PTI and ETI signaling pathways. These three hormones are crucial to effectively defend against plant pathogens. In the case of SAR, which is activated after initial infection with pathogenic strains of microbes, the concentration of SA and the expression of PRs increase even in the distal tissues (Khan et al. 2021). The second type is ISR, which activates plant resistance through specific non-pathogenic strains, and its signals require JA and ethylene (Ali et al. 2022; Hernández-Ruiz et al. 2023).

In addition, other plant hormones, such as IAA and abscisic acid (ABA), were initially considered to be crucial in promoting plant growth and enhancing abiotic tolerance, but are now also considered to be important participants in resisting biotic stress. All plant hormone pathways are connected in a vast and complex network. For example, ethylene, ABA, auxin, gibberellin, and cytokinin are considered important regulators in the SA–JA signaling crosstalk (Tiwari et al. 2021; Hernández-Ruiz et al. 2023). Stomata are the main entry points for pathogens. When pathogens infect plants, they quickly synthesize ABA and initiate a series of defense responses, including upregulating defense-related genes and regulating the stomatal size, thereby inhibiting the invasion and development of pathogens and achieving the goal of protecting plants. At present, studies have found that melatonin is involved in the stomatal behavior of ABA metabolism in tomato (Kou et al. 2021). Melatonin induces stomatal closure in the ABA signaling pathway, hindering pathogen entry. In addition, melatonin also has a similar function in the *CAND2/PMTR1* signaling pathway (Wei et al. 2018a).

11.6 Conclusions and Future Perspectives

More and more evidence shows that melatonin, as a signal molecule, participates in the disease resistance signaling pathway and regulates plant response to biotic stress. Melatonin regulates the generation and reduction of reactive oxygen species in the process of biological stress resistance, which is the first line of defense for melatonin to resist stress. Melatonin induces upregulation of defense genes, rapidly activating the antioxidant defense system, stimulating plant secondary metabolism, and inducing the synthesis of other hormones, which is the second line of defense of melatonin against biological stress. Through these two lines of defense, melatonin enhances the tolerance of plants to biotic stress such as fungi, bacteria, viruses and insects, and weakens the negative impact of biotic stress on plants. These findings provide new ideas for improving crop yield and protecting crops from biotic stress. Melatonin is green, environmentally friendly, safe, and pollution-free, which is beneficial to human health and the environment. The use of melatonin as exogenous treatment in the plant or increasing the content of endogenous melatonin through molecular breeding can improve crop stress resistance, improve crop quality, and enable plants to have better regulatory ability throughout their entire growth period. Therefore, the application of melatonin in crop production is a frontier field worth actively exploring.

Looking ahead, efforts are still needed to explore the following: (1) in-depth research on the metabolic process of melatonin; (2) systematically study the interaction between melatonin and other hormones so as to better improve the molecular regulatory network of melatonin and plant hormones in biotic stress; (3) the specific process of plant perception of melatonin signal, and how to regulate plant tolerance to biotic stress through downstream signal transduction; and (4) exploring the application of melatonin in crop genetic improvement. These will be important hotspots for future research on melatonin, and through more in-depth research, melatonin biology in biotic stress tolerance can be further explored, which can play a greater role in plant disease management in the future.

Acknowledgments This research was supported by the National Key Research and Development Program of China (2018YFD1000800), National Natural Science Foundation of China (31872092, 31872157), Natural Science Foundation of Henan (202300410152, 222102110078), Innovative Research Team (Science and Technology) in University of Henan Province (23IRTSTHN024), and Luoyang Rural Revitalization Project (2101101A).

References

Aghdam MS, Fard JR (2017) Melatonin treatment attenuates postharvest decay and maintains nutritional quality of strawberry fruits (*Fragaria*× *anannasa* cv. Selva) by enhancing GABA shunt activity. Food Chem 221:1650–1657

Ahammed GJ, Li X (2022) Melatonin-induced detoxification of organic pollutants and alleviation of phytotoxicity in selected horticultural crops. Horticulturae 8:1142

Ahammed GJ, Mao Q, Yan Y, Wu M, Wang Y, Ren J, Guo P, Liu A, Chen S (2020) Role of melatonin in arbuscular mycorrhizal fungi-induced resistance to Fusarium wilt in cucumber. Phytopathology 110:999–1009

Ali M, Tumbeh Lamin-Samu A, Muhammad I, Farghal M, Khattak AM, Jan I, ul Haq S, Khan A, Gong Z-H, Lu G (2020) Melatonin mitigates the infection of *Colletotrichum gloeosporioides* via modulation of the chitinase gene and antioxidant activity in *Capsicum annuum* L. Antioxidants 10:7

Ali Q, Zheng H, Rao MJ, Ali M, Hussain A, Saleem MH, Nehela Y, Sohail MA, Ahmed AM, Kubar KA (2022) Advances, limitations, and prospects of biosensing technology for detecting phytopathogenic bacteria. Chemosphere 296:133773

Arnao MB, Hernández-Ruiz J (2015) Functions of melatonin in plants: a review. J Pineal Res 59: 133–150
Arnao MB, Hernández-Ruiz J (2017) Melatonin and its relationship to plant hormones. Ann Bot 121:195–207
Arnao MB, Hernández-Ruiz J (2018) Phytomelatonin versus synthetic melatonin in cancer treatments. Biomed Res Clin Pract 3:1–6
Arnao MB, Hernández-Ruiz J (2019) Melatonin and reactive oxygen and nitrogen species: a model for the plant redox network. Melatonin Res 2:152–168
Chen X, Sun C, Laborda P, Zhao Y, Palmer I, Fu ZQ, Qiu J, Liu F (2018) Melatonin treatment inhibits the growth of *Xanthomonas oryzae* pv. *oryzae*. Front Microbiol 9:2280
Chen L, Wang MR, Li JW, Feng CH, Cui ZH, Zhao L, Wang QC (2019a) Exogenous application of melatonin improves eradication of apple stem grooving virus from the infected in vitro shoots by shoot tip culture. Plant Pathol 68:997–1006
Chen X, Sun C, Laborda P, He Y, Zhao Y, Li C, Liu F (2019b) Melatonin treatments reduce the pathogenicity and inhibit the growth of *Xanthomonas oryzae* pv. *oryzicola*. Plant Pathol 68: 288–296
Hernández-Ruiz J, Giraldo-Acosta M, El Mihyaoui A, Cano A, Arnao MB (2023) Melatonin as a possible natural anti-viral compound in plant biocontrol. Plan Theory 12:781
Khan M, Al Azawi TNI, Pande A, Mun B-G, Lee D-S, Hussain A, Lee B-H, Yun B-W (2021) The role of nitric oxide-induced *ATILL6* in growth and disease resistance in *Arabidopsis thaliana*. Front Plant Sci 12:685156
Kou X, Zhou J, Wu CE, Yang S, Liu Y, Chai L, Xue Z (2021) The interplay between ABA/ethylene and NAC TFs in tomato fruit ripening: a review. Plant Mol Biol 106:223–238
Lee HY, Back K (2016) Mitogen-activated protein kinase pathways are required for melatonin-mediated defense responses in plants. J Pineal Res 60:327–335
Lee HY, Byeon Y, Tan DX, Reiter RJ, Back K (2015) *Arabidopsis* serotonin *N*-acetyltransferase knockout mutant plants exhibit decreased melatonin and salicylic acid levels resulting in susceptibility to an avirulent pathogen. J Pineal Res 58:291–299
Lerner AB, Case JD, Takahashi Y, Lee TH, Mori W (1958) Isolation of melatonin, the pineal gland factor that lightens melanocyteS[1]. J Am Chem Soc 80:2587–2587
Li C, He Q, Zhang F, Yu J, Li C, Zhao T, Zhang Y, Xie Q, Su B, Mei L (2019) Melatonin enhances cotton immunity to *Verticillium* wilt via manipulating lignin and gossypol biosynthesis. Plant J 100:784–800
Li Z, Zhang S, Xue J, Mu B, Song H, Liu Y (2022) Exogenous melatonin treatment induces disease resistance against *Botrytis cinerea* on post-harvest grapes by activating defence responses. Foods 11:2231
Liu C, Chen L, Zhao R, Li R, Zhang S, Yu W, Sheng J, Shen L (2019) Melatonin induces disease resistance to *Botrytis cinerea* in tomato fruit by activating jasmonic acid signaling pathway. J Agric Food Chem 67:6116–6124
Lu R, Liu Z, Shao Y, Sun F, Zhang Y, Cui J, Zhou Y, Shen W, Zhou T (2019) Melatonin is responsible for rice resistance to rice stripe virus infection through a nitric oxide-dependent pathway. Virol J 16:1–8
Mandal MK, Suren H, Ward B, Boroujerdi A, Kousik C (2018) Differential roles of melatonin in plant-host resistance and pathogen suppression in cucurbits. J Pineal Res 65:e12505
Moustafa-Farag M, Almoneafy A, Mahmoud A, Elkelish A, Arnao MB, Li L, Ai S (2019) Melatonin and its protective role against biotic stress impacts on plants. Biomol Ther 10:54
Murch S, KrishnaRaj S, Saxena P (2000) Tryptophan is a precursor for melatonin and serotonin biosynthesis in in vitro regenerated St. John's wort (*Hypericum perforatum* L. cv. Anthos) plants. Plant Cell Rep 19:698–704
Nawaz K, Chaudhary R, Sarwar A, Ahmad B, Gul A, Hano C, Abbasi BH, Anjum S (2020) Melatonin as master regulator in plant growth, development and stress alleviator for sustainable agricultural production: current status and future perspectives. Sustainability 13:294

Nehela Y, Killiny N (2018) Infection with phytopathogenic bacterium inhibits melatonin biosynthesis, decreases longevity of its vector, and suppresses the free radical-defense. J Pineal Res 65: e12511

Qian Y, Tan D-X, Reiter RJ, Shi H (2015) Comparative metabolomic analysis highlights the involvement of sugars and glycerol in melatonin-mediated innate immunity against bacterial pathogen in *Arabidopsis*. Sci Rep 5:15815

Sharif R, Mujtaba M, Ur Rahman M, Shalmani A, Ahmad H, Anwar T, Tianchan D, Wang X (2018a) The multifunctional role of chitosan in horticultural crops; a review. Molecules 23:872

Sharif R, Xie C, Zhang H, Arnao MB, Ali M, Ali Q, Muhammad I, Shalmani A, Nawaz MA, Chen P (2018b) Melatonin and its effects on plant systems. Molecules 23:2352

Shi H, Chen Y, Tan DX, Reiter RJ, Chan Z, He C (2015) Melatonin induces nitric oxide and the potential mechanisms relate to innate immunity against bacterial pathogen infection in *Arabidopsis*. J Pineal Res 59:102–108

Sofy AR, Sofy MR, Hmed AA, Dawoud RA, Refaey EE, Mohamed HI, El-Dougdoug NK (2021) Molecular characterization of the *Alfalfa mosaic virus* infecting *Solanum melongena* in Egypt and the control of its deleterious effects with melatonin and salicylic acid. Plan Theory 10:459

Song W, Forderer A, Yu D, Chai J (2021) Structural biology of plant defence. New Phytol 229: 692–711

Sun Y, Liu Z, Lan G, Jiao C, Sun Y (2019) Effect ofexogenous melatonin on resistance of cucumber to downy mildew. Sci Hortic 255:231–241

Tiwari RK, Lal MK, Kumar R, Mangal V, Altaf MA, Sharma S, Singh B, Kumar M (2021) Insight into melatonin-mediated response and signaling in the regulation of plant defense under biotic stress. Plant Mol Biol 109(4–5):385–399

Vantassel D, Roberts N, Oenill S (1995) Melatonin from higher-plants-isolation and identification of *N*-acetyl 5-methoxytryptamine. Plant Physiol 20855:101–101

Wang R, Yang X, Xu H, Li T (2016) Research progress of melatonin biosynthesis and metabolism in higher plants. Plant Physiol J 52:615–627

Wang L, Feng C, Zheng X, Guo Y, Zhou F, Shan D, Liu X, Kong J (2017) Plant mitochondria synthesize melatonin and enhance the tolerance of plants to drought stress. J Pineal Res 63: e12429

Weeda S, Zhang N, Zhao X, Ndip G, Guo Y, Buck GA, Fu C, Ren S (2014) *Arabidopsis* transcriptome analysis reveals key roles of melatonin in plant defense systems. PLoS One 9: e93462

Wei Y, Hu W, Wang Q, Zeng H, Li X, Yan Y, Reiter RJ, He C, Shi H (2017) Identification, transcriptional and functional analysis of heat-shock protein 90s in banana (*Musa acuminata* L.) highlight their novel role in melatonin-mediated plant response to F usarium wilt. J Pineal Res 62:e12367

Wei J, Li DX, Zhang JR, Shan C, Rengel Z, Song ZB, Chen Q (2018a) Phytomelatonin receptor PMTR 1-mediated signaling regulates stomatal closure in *Arabidopsis thaliana*. J Pineal Res 65: e12500

Wei Y, Chang Y, Zeng H, Liu G, He C, Shi H (2018b) *RAV* transcription factors are essential for disease resistance against cassava bacterial blight via activation of melatonin biosynthesis genes. J Pineal Res 64:e12454

Wei Y, Zhu B, Ma G, Shao X, Xie H, Cheng X, Zeng H, Shi H (2022) The coordination of melatonin and anti-bacterial activity by EIL5 underlies ethylene-induced disease resistance in cassava. Plant J 111:683–697

Wu X, Ren J, Huang X, Zheng X, Tian Y, Shi L, Dong P, Li Z (2021) Melatonin: biosynthesis, content, and function in horticultural plants and potential application. Sci Hortic 288:110392

Yin L, Wang P, Li M, Ke X, Li C, Liang D, Wu S, Ma X, Li C, Zou Y (2013) Exogenous melatonin improves *Malus* resistance to Marssonina apple blotch. J Pineal Res 54:426–434

Yuan M, Jiang Z, Bi G, Nomura K, Liu M, Wang Y, Cai B, Zhou J-M, He SY, Xin X-F (2021) Pattern-recognition receptors are required for NLR-mediated plant immunity. Nature 592:105–109

Zeng W, Mostafa S, Lu Z, Jin B (2022) Melatonin-mediated abiotic stress tolerance in plants. Front Plant Sci 13:847175

Zhang S, Zheng X, Reiter RJ, Feng S, Wang Y, Liu S, Jin L, Li Z, Datla R, Ren M (2017) Melatonin attenuates potato late blight by disrupting cell growth, stress tolerance, fungicide susceptibility and homeostasis of gene expression in *Phytophthora infestans*. Front Plant Sci 8:1993

Zhang S, Liu S, Zhang J, Reiter RJ, Wang Y, Qiu D, Luo X, Khalid AR, Wang H, Feng L (2018) Synergistic anti-oomycete effect of melatonin with a biofungicide against oomycetic black shank disease. J Pineal Res 65:e12492

Zhang Z, Wang T, Liu G, Hu M, Yun Z, Duan X, Cai K, Jiang G (2021) Inhibition of downy blight and enhancement of resistance in litchi fruit by postharvest application of melatonin. Food Chem 347:129009

Zhang W, Li J, Huang T, Xia M, Lu J, Xu X, Liu H (2023) Exogenous melatonin mediates radish (*Raphanus sativus*) and *Alternaria brassicae* interaction in a dose-dependent manner. Front Plant Sci 14:667

Zhao D, Yu Y, Shen Y, Liu Q, Zhao Z, Sharma R, Reiter RJ (2019a) Melatonin synthesis and function: evolutionary history in animals and plants. Front Endocrinol 10:249

Zhao L, Chen L, Gu P, Zhan X, Zhang Y, Hou C, Wu Z, Wu YF, Wang QC (2019b) Exogenous application of melatonin improves plant resistance to virus infection. Plant Pathol 68:1287–1295

Zhao D, Wang H, Chen S, Yu D, Reiter RJ (2021) Phytomelatonin: an emerging regulator of plant biotic stress resistance. Trends Plant Sci 26:70–82

Chapter 12
Role of Melatonin on Post-Harvest Physiology of Crops

Vivek S. Mehta

Abstract Melatonin, a versatile chemical best known for controlling sleep-wake cycles in mammals, has recently emerged as an important regulator of crop post-harvest physiology. This abstract presents an overview of current knowledge about melatonin's impact on post-harvest processes, offering insight on its potential applications in increasing shelf life and preserving crop quality. The antioxidative and free-radical scavenging abilities of melatonin have been widely studied in the context of post-harvest physiology. Its capacity to reduce oxidative stress by neutralizing reactive oxygen species (ROS) helps to maintain cellular integrity, reducing the degeneration of harvested crops. Furthermore, melatonin modulates phytohormone balance, including abscisic acid (ABA) and ethylene.

Exogenous melatonin has shown encouraging results in a variety of crops, delaying senescence, reducing rot, and retaining nutritional content. Melatonin treatments have also shown promise in improving chilling and drought tolerance, emphasizing their importance in post-harvest settings. Despite the advances, numerous aspects demand more investigations. The exact processes underpinning melatonin's impacts on post-harvest physiology require further investigation. For application of melatonin during pre and post harvest, dosage optimization, and research into potential interactions with other storage technologies are required. Finally, melatonin emerges as a critical regulator of crop post-harvest physiology, influencing oxidative stress, phytohormone balance, and stress responses. Its ability to lengthen shelf life, maintain quality, and improve stress tolerance bodes well for sustainable agriculture and reduced food waste. It is critical to conduct additional study.

V. S. Mehta (✉)
Regional Horticulture Research Station, Navsari Agricultural University, Navsari, Gujarat, India

A. Sharma, G. J. Ahammed (eds.), *Melatonin in Plants: Role in Plant Growth, Development, and Stress Response*, Plant Life and Environment Dynamics,
https://doi.org/10.1007/978-981-99-8051-2_12

12.1 Introduction

Melatonin is a multifunctional plant molecule which is distributed ubiquitously in across the plant species (Lo'ay and Taher 2018). As been considered as a master regulator, melatonin is involved in many physiological processes in plants such as plant growth and flowering (Zhang et al. 2019), fruit development, seed and root development, fruit ripening, and response to biotic and abiotic stresses (Wang et al. 2016; Sun et al. 2015; Li et al. 2019a, b; Cao et al. 2018a, b; Sharma and Zheng 2019; Sharma et al. 2020). Melatonin (N-acetyl-5-methoxytryptamin) is produced by the shikimate pathway through enzymatic activities of ryptophan decaboxylaze (TDC), tryptamine 5-hydroxylase (T5H), tryptophan hydroxylase (TPH), serotonin N-acetyletransferase (SNAT). N-acetylserotonin methyltransferase (ASMT) and the caffeic acid o-methyltransferase (COMT). Researchers have confirmed the presence of melatonin in various agricultural produces such as fruits, vegetables, flowers, medicinal and aromatic crops. Many of these crops provide essential nutrients, having medicinal properties and important source of foods across the globe (Mahajan et al. 2017). However, improper and unscientific post-harvest management and storage can be detrimental to the quality and shelf life of these perishable commodities. The post-harvest losses have been increasing from the last decades and leads to significant economic losses (Aghdam et al. 2020b). Recently, new and eco-friendly scientific strategies to generate tolerance to various post-harvest deterioration have been explored which can sustain the quality and shelf life during post-harvest management. Melatonin in plants is synthesized from tryptophan derivatives in different physiological situations. Along with prominent reactive oxygen species (ROS) and reactive nitrogen species (RNS) scavenging potential of melatonin it regulates the expression of a great variety of genes in different physiological processes. All these regulatory functions of melatonin depend on its capacity to interactions with plant growth regulators such as gibberellins (GAs), ethylene, abscisic acid (ABA), jasmonic acid (JA), salicylic acid (SA), hydrogen peroxide (H_2O_2), nitric oxide (NO), and hydrogen sulfide (H_2S) during different physiological development of plants (Zhao et al. 2020; Ma et al. 2021a; Mukherjee and Bhatla 2020; Sun et al. 2021; Siddiqui et al. 2021). This chapter provides the comprehensive view of the potential of melatonin in post-harvest physiology of different agricultural commodities to prevent decay and losses.

12.2 Melatonin in Crops and Post-Harvest

12.2.1 Melatonin Synthesis During Post-Harvest

Melatonin (MLT) synthesized in plant parts after harvest such as tissues, peels, pulp, and seeds (Back et al. 2016). Tryptophan carboxylase and catalyzes tryptophan are the first two enzymes involved in biosynthesis of melatonin which activates tryptophan which then forms serotonin (5-hydroxitryptamine) by hydroxylation of tryptamine. The formation of serotonin is the most important step in the melatonin synthesis pathway (Tan et al. 2013). Afterwards three distinctive enzymes serotonin-N-acetyltransferase (SNAT), N-acetylserotonin methyltransferase (ASMT), caffeic acid o-methyltransferase (COMT) involve in the melatonin synthesis from serotonin (Back et al. 2016). However, it has been reported in some studies that the melatonin synthesis is mainly located in the chloroplast, cytoplasm, and endoplasmic reticulum of plants which are devoid mostly in fruit crops during post-harvest (Back et al. 2016). There are numerous studies have been done on the protective role of melatonin in various plant physiological responses under biotic and abiotic stress which these includes seed germination, crop growth and yield, photosynthesis, water availability, osmoregulation, stomatal conductance and its relation with the metabolism of carbohydrates, lipids, macro and micronutrient compounds (Arnao and Hernandez-Ruiz 2018, 2020).

The possible role of melatonin during post-harvest physiology of crops has also been observed in some studies. Here in this chapter Table 12.1 contains the reviewed studies related to the role of melatonin on post-harvest management of horticultural crop is formatted (Arnao and Hernández-Ruiz 2020). Melatonin seems to promotes fruit ripening and retard senescence during post-harvest. It also regulated ethylene biosynthesis and pigment synthesis, regulated various enzymes of the cell wall, carotenoids, flavonoids, anthocyanins and other phenolic compounds by regulating gibberellin levels during fruiting it sometimes induces parthenocarpy. In vegetable crops, melatonin inhibits the expression of chlorophyll degradation and senescence induced genes, ultimately reduce the decaying and improved the quality of leafy vegetables. The major loss during post-harvest of fresh produces occurred by the loss of cell membrane integrity and function which increases membrane leakage (Dumas et al. 2003). This dysfunctioning of cell membrane caused by the excessive production of ROS viz., H_2O_2, O_2^- and OH^- which malfunctions the biological macromolecules and affecting the metabolism of post-harvest physiological functions of fresh produces. The enzymatic antioxidant system is a first step to control ROS production which regulates the peroxidation. Due to the low content of endogenous melatonin in majority of fresh agricultural commodities after harvesting; its necessary to apply exogenous melatonin for improving post-harvest preservation to aid the endogenous melatonin synthesis.

12.3 Ripening and Senescence

In the process of ripening there are two types of commodities are categorized. A climacteric fruit undergo a unique ripening change after harvesting. There are some studies indicate that exogenous application of melatonin during the harvest affects the physiological pathways involved in biosynthesis of ethylene, regulate genomics and proteomic properties of climacteric and non-climacteric fruits (Pang et al. 2020; Arnao and Hernández-Ruiz 2020). There are numerous studies that have proved the

Table 12.1 Melatonin and its effects on crop phonological, physiological and biochemical changes during post-harvest storage

Crops	Phenological changes	Increased physiological and biochemical indexes	Decreased physiological and biochemical indexes	References
Apple	Reduced decay and senescence	MLT, TAC, CAT, PAL, POD, PPO, SOD activities	PPO	Onik et al. (2020)
Banana	Reduced ripening and senescence	MLT	ROS	Li et al. (2019b)
Citrus	Reduced senescence	AsA, MLT, dioxygenase, monooxygenase, peroxidase	SOD, H_2O_2	Wang et al. (2019), Li et al. (2019a, b)
Grape	Reduced decay and senescence	MLT, H_2O_2, flavonoids and phenols	SOD	Gao et al. (2020)
Kiwifruit	Reduced decay and senescence	AsA, flavonoids, GSH, TAC, CAT, SOD	ROS	Hu et al. (2018b)
Mango	Reduced ripening and senescence	AsA, Anthocyanins, CAT, POD, flavonoids, phenols	PPO	Rastegar et al. (2020)
Peach	Reduced chilling injuries, decay and senescence	AsA, CAT, APX, MDHAR, GR, POD	SOD, PPO, LOX	Cao et al. (2018a, b)
Pomogranate	Reduced chilling injury	DPPH-S, phenols, CAT, SOD, GD, PAL	H_2O_2, PPO, LOX	Jannatizadeh et al. (2019)
Strawberry	Reduced decay	MLT, anthocyanin, flavonoids	H2O2, CAT, APX	Liu et al. (2018)
Tomato	Delayed senescence and reduced chilling injury	MLT, NO, anthocyanin, flavonoids, phenols, CHS, CAT, POD, SOD	H_2O_2, LOX, POD	Li et al. (2019a), Azadshahraki et al. (2018)
Cut gardenia flowers	Reduced senescence and increased shelf life	MLT accumulation, SOD, CAT, APX, flavonoids, anthocyanins	H_2O_2, MDA, PPO	Zhang et al. (2018)
Cabbage leaf	Delayed browning	MLT accumulation, chlorophyll accumulation	ABA, PPH	Tan et al. (2019)
Broccoli floret	Reduced yellowing and increased shelf life	MLT, intracellular ATP, AOX pathway, SOD, CAT, chlorophyll accumulation	H_2O_2, $O_2^{\bullet -}$	Zhu et al. (2018)

MLT melatonin, *TAC* total antioxidant capacity, *CAT* catalase, *PAL* phenylalnine ammonialyase, *POD* peroxidase, *PPO* polyphenol oxidase, *SOD* superoxidase dimustase, *ROS* reactive oxygen species, *AsA* ascorbic acid, *GSH* glutathione, *APX* ascorbate peroxidase, *MDHAR* monodehydroascorbate reductase, *GR* glutathione reductase, *LOX* lipoxygenase, *DPPH-S* DPPH scavenging capacity, *MDA* malondialdehyde, *ABA* abscisic acid, *ATP* adenosine tryphosphate

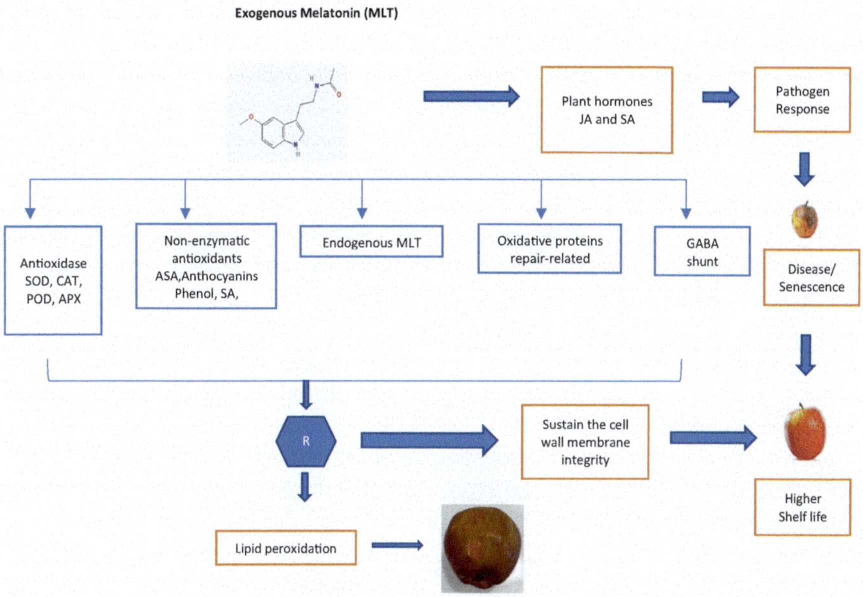

Fig. 12.1 Exogenous melatonin-mediated post-harvest mechanism

effect of melatonin on various post-harvest physiological metabolism of various crops as shown in (Fig. 12.1). Exogenous melatonin treatment have proved to delay some post-harvest attributes such as ripening and color development in fruits such as mango (Liu et al. 2020b), kiwifruit (Hu et al. 2018a), guava (Fan et al. 2022), banana (Hu et al. 2017), pear (Liu et al. 2019), banana (Hu et al. 2017) due to reduce or restraining the starch and cell wall degradation enzymes. Along with this other quality attributes such as acidity, soluble sugar, volatile compounds, respiratory rate, carotenoids, and organic acid are also seeming to be maintained with the melatonin application.

Application of 50 μM melatonin have been observed to stimulate the ripening of tomato fruit (*Solanum lycopersicum* L.) during climacteric peak. The biosynthesis pathway indicated that melatonin activates autocatalytic system of ethylene which affects ACC synthase, ACC oxidase, and ethylene response factor 2 (Sun et al. 2015). These processes stimulate the fruit pigmentation through the accumulation of lycopene by increasing the enzyme activities of phytolene synthase (PYS), acyl carrier protein (ACP), acetyl-CoA carboxylase (ACCase), and carotenoid isomerase (CRTISO) and ultimately boosts up carotenoid pigmentation and fatty acid biosynthesis (Sun et al. 2016). Some studies have also reported that melatonin boosts the expression of enzymes involved in biosynthesis of aromatic compounds in tomato fruit such as hexanal. Application of melatonin increases lycopene accumulation and biosynthesis of aromatic ester compounds during storage (Liu et al. 2019). Shan et al. (2022) and Sun et al. (2020) have noted that melatonin accelerate fruit ripening by stimulating carotenoid biosynthesis which facilitates accumulation of lycopene and alpha-beta-carotene. Hence, melatonin can be a potential molecule to increase the commercial value of tomato fruit during post-harvest quality. Similar results were found by Tijero et al. (2019), they have noted the delayed in ripening of sweet cherry (*Prunus avium* L.). This phenomenon was observed due to cross talk of melatonin with cytokines, jasmonic acid, salicylic acid which results in increased ABA content and reduction in anthocyanins. Delayed in ripening of some fruits such as banana, peach, mango, apple, etc. were observed by many researchers. Liu et al. (2020b) have observed that melatonin delays the mango ripening by plummeting the enzymes such as cell wall pectinase, reduction in endogenous ABA, depolymerization and dimethyl esterification in mango. In jujube fruit (*Ziziphus jujube* Mill.), melatonin has shown to delayed the softening by inhibiting the activities of cellulose, b-glucosidase, PG and pectin methylesterase enzymes (Lauxmann et al. 2014).

Many studies have evaluated the effect of melatonin on ethylene production during storage. Xu et al. (2018) have noted that melatonin accelerated ripening of grape berry (*Vitis vinifera* L.) by boosting ethylene production, anthocyanin levels, increasing ABA accumulation, and H_2O_2 signaling despite being a non-climacteric fruit. Similarly, accelerated ripening was observed in strawberry (Fragaria X ananassa cv. Sabrina) accumulation of ABA and fruit pigmentation through phenylpropanoid pathway and sucrose accumulation (Mansouri et al. 2021; Vallarino et al. 2015).

Due to relatively short period of storage life, vegetable crops have more perishable nature (Mythili and Sathiavelu 2010). Many studies have indicated the role of melatonin in post-harvest cycle of vegetable crops. Xin et al. (2017) have indicated that 100 μM melatonin solution immersion of cucumber for 2 h delay senescence and extend shelf life. Soaking of broccoli in 100 μM melatonin solution for 5 min extended the shelf life for 5–7 days due to increase in mitochondrial AOX capacity (Zhu et al. 2018). This leads to increased ATP availability, antioxidant SOD, and catalase (CAT) and thereby reduction in endogenous ROS which delays the chlorophyll degradation. Some studies have reported that the content of y-aminobutyric acid (GABA) significantly increased in strawberry and pear after treated with melatonin (Rastegar et al. 2020; Sheng et al. 2017). As a tryptophan derivative, melatonin has found to be induced the IAA accumulation and increase the shelf life of banana during storage (Li et al. 2019b). Similarly, Wu et al. (2021) have noted that fresh broccoli treated with Melatonin stored at 20 °C for 4 days have reduced the yellowing of florets by reducing the activity of enzymes chlorophyllase (Chlase), phenophytinase (PPH), and Pheide-a-oxygenase (PaO). Cassava roots treated with melatonin have shown significant reduction in H_2O_2 content, delayed post-harvest physiological deterioration (Hu et al. 2016; Ma et al. 2016).

Perishable commodities such as fruits and vegetables gradually deteriorate during the ripening process which termed as a senescence. The exogenous melatonin treatment significantly reduces the senescence during post-harvest. The senescence of fresh produce is accompanied by loss in cell wall integrity which is triggered by increase in membrane leakage (Dumas et al. 2003). During post-harvest storage of fresh commodities, activities of ROS increases continuously due to various stress factors which induces lipid peroxidation, which catalyze the oxygenation of unsaturated fatty acids through ROS and lipid peroxidase (Li et al. 2016). Melatonin also impacts on storage life by affecting the senescence mechanism in which ROS play an important role. To scavenge ROS, fruits and vegetables have developed efficient antioxidant mechanism to protect against oxidative damage during storage (Jiang et al. 2004). Subsequently these molecules are decomposed to produce oxidative free radicles. As given in Table 12.1 melatonin application could significantly triggered the antioxidant mechanism by activating enzymes such as, ascorbate peroxidase (APX), dehydroascorbate reductase (DHAR), glutathione peroxidase (GPX), catalase (CAT), dehydrothione reductase (GR), glutathione s-transferase (GST), monodehydroascorbate reductase (MDHAR), peroxidase (POD), polyphenol oxidase (PPO), and superoxide dismutase (SOD) (Yun et al. 2021), with non-enzymatic antioxidant mainly composed of flavonoids, phenols, carotenoids, anthocyanins, ascorbic acid (AsA), dehydroascorbic acid (DHA), and glutathione (GSH) which are responsible for ROS homeostasis to check the stress which reduces the senescence and increase shelf life of perishable produces. Tian et al. (2013) have noted that melatonin increases the levels of the antioxidant enzymes, catalase, SOD and ascorbate-glutathione cycle components and reduce the level of NADPH oxidase and lipoxygenase resulted in regulation of ROS metabolism there by reduced the cell degradation. An initial study performed by Arnao and Hernández-Ruiz (2009) revealed that melatonin application delayed the dark induced senescence in barley leaves. In peach fruit, 100 μM melatonin solution treatment for 10 min have delayed the fruit senescence and flesh browning via melatonin-mediated regulation on membrane fatty acid metabolism, phenolic metabolism, and ROS system (Gao et al. 2018). Melatonin have noted to effectively inhibit the expression of genes related to ethylene biosynthesis such as Musa 1-aminocyclopropane-1-carboxylate oxidase 1 and Musa 1-aminocyclopropane-1-carboxylate synthase 1 which delayed the post-harvest senescence of banana (Hu et al. 2017). Melatonin application in Lychee (*Litchi chinensis*) during storage have inhibited the pericarp browning, reduction in cell membrane leakage, delay discoloration along with regulating the production of superoxide anions, hydrogen peroxide, and malondialdehyde (Zhang et al. 2018). Sweet cherry treated with melatonin during storage have also shown the delay in senescence and retained the quality. Melatonin have raised phenol, flavonoids, and anthocyanin levels following by upscaling the activities of PAL and CHS, reduced polyphenol oxidase (PPO) which leads to DPPH scavenging capacity, higher SOD, ascorbate peroxidase (APX), CAT, and glutathione reductase (GR). On the contrary, MEL reduced the enzymatic activities of phospholipase D (PLD) and LOX. These both physiological metabolism changes maintain membrane integrity by lowering endogenous H_2O_2 and malondialdehyde (MDA) accumulation (Sharafi et al. 2021).

12.4 Melatonin and Post-Harvest Diseases

Post-harvest storage of fresh produces are often accompanied by fungal/bacterial infections which leads to rapid senescence. Defense mechanism is crucial for post-harvest disease resistance. Li et al. (2019a) have reported that tomato fruits treated with external melatonin application revealed increase in the content of NO, H_2O_2, and O_2^-. Melatonin and its coordination with NO and ROS have been found to trigger disease resistance in fresh produces (Liu et al. 2019; Zhang et al. 2020). These induced phytohormones and signaling molecules viz., salicylic acid (SA), jasmonic acid (JA) and mitogen-activated protein kinase (MAPK) signaling pathway which induces the disease tolerance (Tijero et al. 2019). The phenylpropanoid pathway contributes tolerance by providing SA which triggers a systemic acquired response (SAR) to pathogens (Lefevere et al. 2020). Tomato fruits treated with MLT was found to alleviate the gray mold decay (*B. cinerea*) through the accumulation of SA which upscaled the activity of enzymes involved in the phenylpropanoid pathway, pathogenesis-related genes (*PR*) thereby increasing chitinase and beta-glucanase activity (Li et al. 2019a). Similarly, exogenous application of melatonin inhibits the fungal lesions of Litchi fruit during storage with NADPH accumulation by triggering phenolic and flavonoid accumulation, increase enzymatic activity of PAL, cinnamate-4-hydroxylase (C_4H), mitochondrial metabolisms (SDH, CCO, and H+-ATPase), and oxidative pentose phosphate pathway (OxPP) (Zhang et al. 2021b). Chen et al. (2020) have found that melatonin application provides resistant to fungal decay during post-harvest storage of Wax apple. Melatonin acts by reducing the LOX activity which preserve the cell membrane integrity and increases the level of GABA pathway. Fresh potato tubers slices treated with melatonin have revealed the size reduction in lesions infected by post-harvest potato blight (Zhang et al. 2017). The application of melatonin (200 μM) for 72 h in apple have reduced the infestation of gray mold (Botrytis cinerea) during storage (Cao et al. 2017). These studies have confirmed that melatonin might play roles in delaying senescence of post-harvest horticultural commodities by the involvement in both IAA and ABA mechanisms. In general, melatonin has the potential to be used as an anti-senescence molecule to reduce decay and increase shelf life of perishable horticultural produces during post-harvest.

12.5 Melatonin and Post-Harvest Chilling Injury

As a post-harvest shelf life prolong technology, low temperature storage is of widely used technology (Azadshahraki et al. 2018). However, during storage chilling injuries in the fresh produces may occur and reduce the flavor and quality (Valenzuela et al. 2017). Chilling injury is mainly associate with the overproduction of ROS (Gill and Tuteja 2010; Zeng et al. 2016). Chilling shock is first perceived by membranes, where transition of fluid liquid crystalline phase to solid gel phase happens. Afterwards, the higher electrolyte leakage occurs which leads to membrane malfunctions in proteins and prevents intracellular ATP transportations and ATP shortage. Due to disruption in Ca^{+2}-ATPase activity, there is an increase in ROS by rising electrolyte leakage from NADH dehydrogenase, cytochrome-c-oxidase activity accompanied by cytosolic Ca^{+2} production (Sevillano et al. 2009). During this, higher binding capacity of Ca^{+2} to PLD C2, PLD enzyme shows capacity to bind to phospholipids with a negative charge, the reason behind as choline and ethanolamine carry a positive charge (Pak Dek et al. 2020). During senescence and stress such as low temperature exposure, PI3K activity upscales which triggers the ethylene signal transduction pathway and increase PLD activity which initiates membrane degradation and production of free linoleic acids (Padmanabhan and Paliyath 2018).

Melatonin can play a pivotal role to reduce the chilling injuries by the reduction of malondialdehyde (MDA), electrolyte leakage, and lipoxygenase (LOX) activities during post-harvest. Melatonin preserves the membrane integrity by reduction in O_2 and H_2O_2 and increase in phenol, anthocyanin productions which lowers the electrolyte leakage and MDA. Kong et al. (2020) have found that bell pepper (*Capsicum annuum* L.) treated with 100 μM MT reduced the chilling injury by triggering the antioxidant mechanism and reducing cold induced membrane lipid peroxidation during cold storage at 4 °C. Litchi fruits treated with 400 μM MT dipping treatment showed longed shelf life by reduction in the MDA content and chilling injury, inhibition in discoloration with keeping membrane integrity intact (Liu et al. 2020a). Post-harvest physiological deterioration in Cassava roots was found to be delay with increase in MLT during storage, which acts by its ROS scavenging capacity to protect membrane (Hu et al. 2016). Mushroom cap browning during storage was found to be reduced with exogenous MLT application. MLT treatment increased the NADPH oxidase activity to reduce the level of H_2O_2. This activation provides tryptophan for MLT production internally which reflected by increment in shikimate dehydrogenase enzyme activities (Aghdam et al. 2019b). Similarly, reduction in spathe browning was observed in anthurium (*Anthurium andraeanum*) cut flowers which were treated with stem dipping in MLT for 15 min at 20 °C and then stored at 4 °C for 21 days. Melatonin triggered the ROS metabolism and increased the SOD, CAT, APX, and GR activities thereby prevents MDA accumulations (Aghdam et al. 2019a) Table 12.2 formulated the numerous studies on beneficial effect of MLT during storage in terms of chilling injuries (Sati et al. 2022).

Melatonin and its synergy with nitric oxide (NO) and H_2O_2 induced enzymatic and non-enzymatic antioxidants in post-harvest produces to scavenge the ROS and RNS which may induced during cold storage by heat shock proteins and increase in the contents of salicylic acid, unsaturated fatty acids and nitrogenous compounds, which triggers the cold-tolerance mechanism.

Table 12.2 MLT application and its impact on chilling injury index during storage

Crops	Melatonin concentration applied during storage (μM)	Chilling injury index	References
Apricot	100	↓ 22.31% (28 days storage)	Medina-Santamarina et al. (2021)
Banana	200	↓ 28.57% (23 days storage)	Wang et al. (2021)
Bell pepper	100	↓ 22.22% (23 days storage)	Kong et al. (2020)
Cucumber	100	↓ 37.5% (15 days storage)	Madebo et al. (2021)
Mango	100	↓ 80% in Langra variety (28 days of storage)	Bhardwaj et al. (2021)
Nectarine	1000	↓ 68% (7 days of storage)	Bal (2021)
Peach	100	↓ 73.33% (28 days of storage)	Cao et al. (2016)
Pomegranate	100	↓ 28% (123 days of storage)	Molla et al. (2022)
Sapota	90	↓ 27.78% (30 days of storage)	Mirshekari et al. (2020)

↓ indicate the lowest percentage of damage

12.6 Melatonin and Post-Harvest Nutritional Quality

Recently, consumers have been targeting attention to the nutritional quality of fresh agricultural produces as these produces provides macro and micronutrients to the consumers having numerous beneficial health effects. It has been studied that exogenous application of melatonin can play a pivotal role in preserving the nutritional quality of fresh commodities during post-harvest storage (Aghdam et al. 2020a; Aghdam and Fard 2017; Liu et al. 2018). Pre-harvest application of melatonin in grape berry boosts the ethylene production by activating ACS1 which increases flavonoid, phenolic acid, stilbene by triggering *STS1*, flavanone 3-hydroxylase (F3H), and leucoanthocyanidin reductase (LAR) (Ma et al. 2021b). Currently, the use of chitosan as a safe polysaccharide coating of fresh produces has increased due to its eco-friendly nature for shelf life. Addition of MLT in these packaging films providing potential antioxidant capacity is being seen as a considerable technology to sustain nutritional quality and reduce post-harvest losses (Zhao et al. 2020). Cucumber, broccoli, and melons coated with a chitosan based MLT coating containing 1.2% chitosan, 0.8% carboxymethyl cellulose, and 50 mg/L MLT imparts reduction in weight loss and browning, higher firmness, increased chlorophyll accumulation, higher total soluble solids, and titratable acidity. It also imparts the DPPH scavenging capacity and antibacterial property to the coating (Zhao et al. 2020). Exogenous application of melatonin in jujube fruits boosts enhanced glutathione-S-transferase (GST) activity which effect the endogenous glyphosate, malathion and chlorothalonil degradation which may can get consumed by the end consumers. MLT also exhibit upscaling of ascorbic acid (AA), FRAP scavenging activity, and phenol accumulation (Deng et al. 2021).

12.7 Conclusion

Prominent role of melatonin in the post-harvest physiology have clearly proven the many studies. Melatonin a pivotal signaling molecule in post-harvest metabolism can be increased through exogenous applications which can regulates the levels of ROS, NO and plant hormones which involves in extend shelf life and maintain quality of harvested fresh produces. During post-harvest application melatonin have found to decreases the oxidative damage, degradation of starch and cell wall, synthesis, and release of volatile aroma while it regulates the energy metabolism and expression of enzymatic activities. It also coordinates with IAA, ABA, GABA, ethylene, JA, NO, ROS and SA wo regulate the ripening, senescence, chilling injury and decay, post-harvest disease of agricultural produces. To conclude, the better understanding of the molecular, biochemical, and physiological activities of melatonin during pre- and post-harvest stages can help to develop effective management technology to sustain the quality and increase the storage life of post-harvest fresh commodities during storage.

References

Aghdam MS, Fard JR (2017) Melatonin treatment attenuates postharvest decay and maintains nutritional quality of strawberry fruits (Fragaria×anannasa cv. Selva) by enhancing GABA shunt activity. Food Chem 221:1650–1657

Aghdam MS, Jannatizadeh A, Nojadeh MS, Ebrahimzadeh A (2019a) Exogenous melatonin ameliorates chilling injury in cut anthurium flowers during low temperature storage. Postharvest Biol Technol 148:184–191

Aghdam MS, Luo Z, Jannatizadeh A, Farmani B (2019b) Exogenous adenosine triphosphate application retards cap browning in Agaricus bisporus during low temperature storage. Food Chem 293:285–290

Aghdam MS, Luo Z, Li L, Jannatizadeh A, Fard JR, Pirzad F (2020a) Melatonin treatment maintains nutraceutical properties of pomegranate fruits during cold storage. Food Chem 303: 125385

Aghdam MS, Palma JM, Corpas FJ (2020b) NADPH as a quality foot-printing in horticultural crops marketability. Trends Food Sci Technol 103:152–161

Arnao MB, Hernandez-Ruiz J (2009) Protective effect of melatonin against chlorophyll degradation during the senescence of barley leaves. J Pineal Res 46:58–63. https://doi.org/10.1111/j.1600-079X.2008.00625

Arnao MB, Hernandez-Ruiz J (2018) Melatonin and its relationship to plant hormones. Ann Bot 121:195–207

Arnao M, Hernández-Ruiz J (2020) Melatonin in flowering, fruit set and fruit ripening. Plant Rep 33:77–87

Azadshahraki F, Jamshidi B, Mohebbi S (2018) Postharvest melatonin treatment reduces chilling injury and enhances antioxidant capacity of tomato fruit during cold storage. Adv Hortic Sci 32: 299–309

Back K, Tan DX, Reiter RJ (2016) Melatonin biosynthesis in plants: multiple pathways catalyze tryptophan to melatonin in the cytoplasm or chloroplasts. J Pineal Res 61:426–437

Bal E (2021) Effect of melatonin treatments on biochemical quality and postharvest life of nectarines. J Food Meas Charact 15:288–295

Bhardwaj R, Pareek S, Domínguez-Avila JA, Gonzalez-Aguilar GA (2021) Changes in the activity of proline-metabolising enzymes is associated with increased cultivar-dependent chilling tolerance in mangos, in response to pre-storage melatonin application. Postharvest Biol Technol 182: 11702

Cao S, Song C, Shao J, Bian K, Chen W, Yang Z (2016) Exogenous melatonin treatment increases chilling tolerance and induces defense response in harvested peach fruit during cold storage. J Agric Food Chem 64(25):5215–5222

Cao JJ, Yu ZC, Zhang Y, Li BH, Liang WX, Wang CX (2017) Control efficiency of exogenous melatonin against post-harvest apple grey mold and its influence on the activity of defensive enzymes. Plant Physiol J 53:1753–1760

Cao S, Bian K, Shi L, Chung H-H, Chen W, Yang Z (2018a) Role of melatonin in cell-wall disassembly and chilling tolerance in cold-stored peach fruit. J Agric Food Chem 66:5663–5670

Cao S, Shao J, Shi L, Xu L, Shen Z, Chen W et al (2018b) Melatonin increases chilling tolerance in postharvest peach fruit by alleviating oxidative damage. Sci Rep 8:806

Chen Y, Zhang Y, Nawaz G, Zhao C, Li Y, Dong T et al (2020) Exogenous melatonin attenuates post-harvest decay by increasing antioxidant activity in wax apple (Syzygium samarangense). Front Plant Sci 11:569779

Deng B, Xia C, Tian S, Shi H (2021) Melatonin reduces pesticide residue, delays senescence, and improves antioxidant nutrient accumulation in postharvest jujube fruit. Postharvest Biol Technol 173:111419

Dumas Y, Dadomo M, Di Lucca G, Grolier P (2003) Effects of environmental factors and agricultural techniques on antioxidant content of tomatoes. J Sci Food Agr 83(5):369–382

Fan Y, Li C, Li Y, Huang R, Guo M, Liu J, Ge Y (2022) Postharvest melatonin dipping maintains quality of apples by mediating sucrose metabolism. Plant Physiol Biochem 174:43–50. https://doi.org/10.1016/j.plaphy.2022.01.034

Gao H, Lu ZM, Yang Y, Wang DN, Yang T, Cao MM et al (2018) Melatonin treatment reduces chilling injury in peach fruit through its regulation of membrane fatty acid contents and phenolic metabolism. Food Chem 245:659–666

Gao S, Ma W, Lyu X, Cao X, Yao Y (2020) Melatonin may increase disease resistance and flavonoid biosynthesis through effects on DNA methylation and gene expression in grape berries. BMC Plant Biol 20:231

Gill SS, Tuteja N (2010) Reactive oxygen species and antioxidant machinery in abiotic stress tolerance in crop plants. Plant Physiol Biochem 48:909–930

Hu W, Kong H, Guo Y, Zhang Y, Ding Z, Tie W et al (2016) Comparative physiological and transcriptomic analyses reveal the actions of melatonin in the delay of postharvest physiological deterioration of cassava. Front Plant Sci 7:736

Hu W, Yang H, Tie W, Yan Y, Ding Z, Liu Y et al (2017) Natural variation in banana varieties highlights the role of melatonin in postharvest ripening and quality. J Agric Food Chem 65: 9987–9994

Hu M, Li J, Rao J (2018a) Effect of melatonin on ripening and senescence of postharvest kiwifruits. Food Sci 39:226–232

Hu W, Tie W, Ou W, Yan Y, Kong H, Zuo J et al (2018b) Crosstalk between calcium and melatonin affects postharvest physiological deterioration and quality loss in cassava. Postharvest Biol Technol 140:42–49

Jannatizadeh A, Aghdam MS, Luo Z, Razavi F (2019) Impact of exogenous melatonin application on chilling injury in tomato fruits during cold storage. Food Bioprocess Technol 12:741–750

Jiang Y, Duan X, Joyce D, Zhang Z, Li J (2004) Advances in understanding of enzymatic browning in harvested litchi fruit. Food Chem 88:443–446

Kong XM, Ge WY, Wei BD, Zhou Q, Zhou X, Zhao YB, Ji SJ (2020) Melatonin ameliorates chilling injury in green bell peppers during storage by regulating membrane lipid metabolism and antioxidant capacity. Postharvest Biol Technol 170:111315

Lauxmann MA, Borsani J, Osorio S et al (2014) Deciphering the metabolic pathways influencing heat and cold responses during post-harvest physiology of peach fruit. Plant Cell Environ 37: 601–616

Lefevere H, Bauters L, Gheysen G (2020) Salicylic acid biosynthesis in plants. Front Plant Sci 11: 338

Li H, Dong Y, Chang J, He J, Chen H, Liu Q et al (2016) High-throughput microRNA and mRNA sequencing reveals that microRNAs may be involved in melatonin-mediated cold tolerance in Citrullus lanatus L. Front Plant Sci 7:1231

Li S, Xu Y, Bi Y, Zhang B, Shen S, Jiang T et al (2019a) Melatonin treatment inhibits gray mold and induces disease resistance in cherry tomato fruit during postharvest. Postharvest Biol Technol 157:110962

Li T, Wu Q, Zhu H, Zhou Y, Jiang Y, Gao H et al (2019b) Comparative transcriptomic and metabolic analysis reveals the effect of melatonin on delaying anthracnose incidence upon postharvest banana fruit peel. BMC Plant Biol 19:289

Liu C, Zheng H, Sheng K, Liu W, Zheng L (2018) Effects of melatonin treatment on the postharvest quality of strawberry fruit. Postharvest Biol Technol 139:47–55

Liu J, Liu H, Wu T, Zhai R, Yang C, Wang Z et al (2019) Effects of melatonin treatment of postharvest pear fruit on aromatic volatile biosynthesis. Molecules 24:4233

Liu GS, Zhang Y, Yun Z (2020a) Melatonin enhances cold tolerance by regulating energy and proline metabolism in litchi fruit. Foods 9:454

Liu S, Huang H, Huber DJ, Pan Y, Shi X, Zhang Z (2020b) Delay of ripening and softening in 'Guifei' mango fruit by postharvest application of melatonin. Postharvest Biol Technol 163: 111136

Lo'ay AA, Taher MA (2018) Influence of edible coatings chitosan/PVP blending with salicylic acid on biochemical fruit skin browning incidence and shelf life of guava fruits cv. 'Banati'. Sci Hortic 235:424–436

Ma Q, Zhang T, Zhang P, Wang ZY (2016) Melatonin attenuates postharvest physiological deterioration of cassava storage roots. J Pineal Res 60:424–434

Ma W, Xu L, Gao S, Lyu X, Cao X, Yao Y (2021a) Melatonin alters the secondary metabolite profile of grape berry skin by promoting VvMYB14-mediated ethylene biosynthesis. Hortic Res 8:43

Ma W, Xu L, Gao S, Lyu X, Cao X, Yao Y (2021b) Melatonin alters the secondary metabolite profile of grape berry skin by promoting VvMYB14-mediated ethylene biosynthesis. Hortic Res 8:1–15

Madebo MP, Hu S, Zheng Y, Jin P (2021) Mechanisms of chilling tolerance in melatonin treated postharvest fruits and vegetables: a review. J Future Foods 1(2):156–167

Mahajan PV, Caleb OJ, Gil MI, Izumi H, Colelli G, Watkins CB et al (2017) Quality and safety of fresh horticultural commodities: recent advances and future perspectives. Food Packag Shelf Life 14:2–11

Mansouri S, Sarikhani H, Sayyari M, Aghdam MS (2021) Melatonin accelerates strawberry fruit ripening by triggering GAMYB gene expression and promoting ABA accumulation. Sci Hortic 281:109919

Medina-Santamarina J, Zapata PJ, Valverde JM, Valero D, Serrano M, Guillén F (2021) Melatonin treatment of apricot trees leads to maintenance of fruit quality attributes during storage at chilling and non-chilling temperatures. Agron J 11:917

Mirshekari A, Madani B, Yahia EM, Golding JB, Vand SH (2020) Postharvest melatonin treatment reduces chilling injury in sapota fruit. J Sci Food Agric 100:1897–1903

Molla SMH, Rastegar S, Omran VG, Khademi O (2022) Ameliorative effect of melatonin against storage chilling injury in pomegranate husk and arils through promoting the antioxidant system. Sci Hortic 295:110889

Mukherjee S, Bhatla SC (2020) Exogenous melatonin modulates endogenous H_2S homeostasis and L-Cysteine desulfhydrase activity in salt-stressed tomato (Solanum lycopersicum L. var. cherry) seedling cotyledons. J Plant Growth Regul 40:1–13

Mythili RS, Sathiavelu A (2010) Recovery of bacteriocin (nisin) from *Lactococcus lactis* and testing its ability to increase the shelf life of vegetables (carrot and beans). Res J Biol Sci 5:727–730

Onik JC, Wai SC, Li A, Lin Q, Sun Q, Wang Z et al (2020) Melatonin treatment reduces ethylene production and maintains fruit quality in apple during postharvest storage. Food Chem 337: 127753–127753

Padmanabhan P, Paliyath G (2018) Ethylene signal transduction during fruit ripening and senescence. In: Paliyath G, Subramanian J, Lim L-T, Subramanian KS, Handa AK, Mattoo AK (eds) Postharvest biology and nanotechnology. Wiley, Hoboken, pp 53–87

Pak Dek MS, Padmanabhan P, Tiwari K, Todd JF, Paliyath G (2020) Structural and functional characterization of Solanum lycopersicum phosphatidylinositol 3-kinase C_2 domain. Plant Physiol Biochem 148:180–192

Pang L, Wu Y, Pan Y, Ban Z, Li L, Li X (2020) Insights into exogenous melatonin associated with phenylalanine metabolism in postharvest strawberry. Postharvest Biol Technol 168:111244

Rastegar S, Hassanzadeh Khankahdani H, Rahimzadeh M (2020) Effects of melatonin treatment on the biochemical changes and antioxidant enzyme activity of mango fruit during storage. Sci Hortic 259:108835

Sati H, Khandelwal A, Pareek S (2022) Effect of exogenous melatonin in fruit postharvest, crosstalk with hormones, and defense mechanism for oxidative stress management. Food Front 1:29

Sevillano L, Sanchez-Ballesta MT, Romojaro F, Flores FB (2009) Physiological, hormonal and molecular mechanisms regulating chilling injury in horticultural species. Postharvest technologies applied to reduce its impact. J Sci Food Agric 89:555–573

Shan S, Wang Z, Pu H, Duan W, Song H, Li J et al (2022) DNA methylation mediated by melatonin was involved in ethylene signal transmission and ripening of tomato fruit. Sci Hortic 291: 110566

Sharafi Y, Jannatizadeh A, Fard JR, Aghdam MS (2021) Melatonin treatment delays senescence and improves antioxidant potential of sweet cherry fruits during cold storage. Sci Hortic 288: 110304

Sharma A, Zheng B (2019) Melatonin mediated regulation of drought stress: physiological and molecular aspects. Plan Theory 8(7):190

Sharma A, Wang J, Xu D, Tao S, Chong S, Yan D et al (2020) Melatonin regulates the functional components of photosynthesis, antioxidant system, gene expression, and metabolic pathways to induce drought resistance in grafted Carya cathayensis plants. Sci Total Environ 713:136675

Sheng L, Shen D, Luo Y, Sun X, Wang J, Luo T et al (2017) Exogenous gamma-aminobutyric acid treatment affects citrate and amino acid accumulation to improve fruit quality and storage performance of postharvest citrus fruit. Food Chem 216:138–145

Siddiqui MH, Khan MN, Mukherjee S, Basahi RA, Alamri S, Al Amri AA et al (2021) Exogenous melatonin-mediated regulation of K^+/Na^+ transport, H^+-ATPase activity and enzymatic antioxidative defence operate through endogenous hydrogen sulphide signalling in NaCl-stressed tomato seedling roots. Plant Biol 23:797–805

Sun Q, Zhang N, Wang J, Zhang H, Li D, Shi J et al (2015) Melatonin promotes ripening and improves quality of tomato fruit during postharvest life. J Exp Bot 66:657–668

Sun Q, Zhang N, Wang J, Cao Y, Li X, Zhang H et al (2016) A label-free differential proteomics analysis reveals the effect of melatonin on promoting fruit ripening and anthocyanin accumulation upon postharvest in tomato. J Pineal Res 61:138–153

Sun YD, Guo DL, Yang SD, Zhang HC, Wang LL, Yu YH (2020) Melatonin treatment improves the shelf-life and postharvest quality of table grape (Vitis labrusca L. cv. 'Fengzao'). Journal of Berry Research 10:665–676. https://doi.org/10.3233/jbr-200569

Sun Y, Ma C, Kang X, Zhang L, Wang J, Zheng S et al (2021) Hydrogen sulfide and nitric oxide are involved in melatonin-induced salt tolerance in cucumber. Plant Physiol Biochem 167:101–112

Tan DX, Manchester LC, Liu X, Rosales-Corral SA, Acuna-Castroviejo D, Reiter RJ (2013) Mitochondria and chloroplasts as the original sites of melatonin synthesis: a hypothesis related to melatonin's primary function and evolution in eukaryotes. J Pineal Res 54:127–138

Tan XL, Fan ZQ, Kuang JF, Lu WJ, Reiter RJ, Lakshmanan P et al (2019) Melatonin delays leaf senescence of Chinese flowering cabbage by suppressing ABFs-mediated abscisic acid biosynthesis and chlorophyll degradation. J Pineal Res 67:e12570

Tian S, Qin G, Li B (2013) Reactive oxygen species involved in regulating fruit senescence and fungal pathogenicity. Plant Mol Biol 82:593–602

Tijero V, Munoz P, Munne-Bosch S (2019) Melatonin as an inhibitor of sweet cherries ripening in orchard trees. Plant Physiol Biochem 140:88–95

Valenzuela JL, Manzano S, Palma F, Carvajal F, Garrido D, Jamilena M (2017) Oxidative stress associated with chilling injury in immature fruit: postharvest technological and biotechnological solutions. Int J Mol Sci 18:1467

Vallarino JG, Osorio S, Bombarely A, Casañal A, Cruz-Rus E, Sánchez-Sevilla JF et al (2015) Central role of Fa GAMYB in the transition of the strawberry receptacle from development to ripening. New Phytol 208:482–496

Wang C, Yin L-Y, Shi X-Y, Xiao H, Kang K, Liu X-Y, Zhan J-C, Huang W-D (2016) Effect of cultivar, temperature, and environmental conditions on the dynamic change of melatonin in mulberry fruit development and wine fermentation. J Food Sci 81:M958–M967

Wang J, Hong M, Feng Y, He M, Wang R, Yu Z et al (2019) Effects of exogenous melatonin treatment on fruit quality and ethanol metabolism of 'Aiyuan 38' citrus. Food Ferment Ind 45: 147–154

Wang Z, Pu H, Shan S, Zhang P, Li J, Song H, Xu X (2021) Melatonin enhanced chilling tolerance and alleviated peel browning of banana fruit under low temperature storage. Postharvest Biol Technol 179:111571

Wu C, Cao S, Xie K, Chi Z, Wang J, Wang H et al (2021) Melatonin delays yellowing of broccoli during storage by regulating chlorophyll catabolism and maintaining chloroplast ultrastructure. Postharvest Biol Technol 172:111378

Xin DD, Si JJ, Kou L, P. (2017) Post-harvest exogenous melatonin enhances quality and delays the senescence of cucumber. Acta Hortic Sin 44:891–901

Xu L, Yue Q, Bian FE, Zhai H, Yao Y (2018) Melatonin treatment enhances the polyphenol content and antioxidant capacity of red wine. Hortic Plant J 4:144–150

Yun Z, Gao H, Chen X, Chen Z, Zhang Z, Li T et al (2021) Effects of hydrogen water treatment on antioxidant system of litchi fruit during the pericarp browning. Food Chem 336:127618

Zeng L, Wang YH, Ai CX, Zheng JL, Wu CW, Cai R (2016) Effects of β-glucan on ROS production and energy metabolism in yellow croaker (*Pseudosciaena crocea*) under acute hypoxic stress. Fish Physiol Biochem 42(5):1395–1405

Zhang SM, Zheng XZ, Reiter RJ, Feng S, Wang Y, Liu S et al (2017) Melatonin attenuates potato late blight by disrupting cell growth, stress tolerance, fungicide susceptibility and homeostasis of gene expression in *Phytophthora infestans*. Front Plant Sci 8:1993

Zhang YY, Huber DJ, Hu MJ, Jiang GX, Gao ZY, Xu XB et al (2018) Melatonin delays postharvest browning in litchi fruit by enhancing anti-oxidative processes and oxidation repair. J Agric Food Chem 66:7475–7484

Zhang H, Wang L, Shi K, Shan D, Zhu Y, Wang C, Bai Y, Yan T, Zheng X, Kong J (2019) Apple tree flowering is mediated by low level of melatonin under the regulation of seasonal light signal. J Pineal Res 66:12551

Zhang W, Cao J, Fan X, Jiang W (2020) Applications of nitric oxide and melatonin in improving postharvest fruit quality and the separate and crosstalk biochemical mechanisms. Trends Food Sci Technol 99:531–541

Zhang Z, Wang T, Liu G, Hu M, Yun Z, Duan X et al (2021b) Inhibition of downy blight and enhancement of resistance in litchi fruit by postharvest application of melatonin. Food Chem 347:129009

Zhao H, Wang L, Belwal T, Jiang Y, Li D, Xu Y et al (2020) Chitosan-based melatonin bilayer coating for maintaining quality of fresh-cut products. Carbohydr Polym 235:115973

Zhu LL, Hu HL, Luo SF, Wu ZX, Li PX (2018) Melatonin delaying senescence of postharvest broccoli by regulating respiratory metabolism and antioxidant activity. Trans Chinese Soc Agric Eng 34:300–308

Chapter 13
Advances in Tools and Techniques to Quantify Melatonin in Plants and Foodstuff

Md. Kamrul Hasan, Jannat Shopan, Golam Jalal Ahammed, and Jie Zhou

Abstract As an ancient pleiotropic molecule, melatonin (*N*-acetyl-5-methoxytryptamine) regulates numerous cellular, physiological, biochemical, and molecular processes in plants. Although the existence of melatonin has been well documented over 27 years in photosynthetic organisms, research into the phytomelatonin is still limited due to lower endogenous concentrations in complex plant metrics and lack of rapid efficient quantification techniques. Therefore, scientists worldwide showed extensive efforts to develop sensitive detection tools and techniques for future innovative research and human well-being. Among the various analytical methods selective high performance liquid chromatography mass spectrophotometry (HPLC-MS) is largely used for high selectivity, accuracy, robustness, and low limit of detection. However, in recent years micro-fabrication and nanomaterial-mediated decoration of novel electrochemical and optical detection tools draw extra attention in addition to capillary electrophoresis, fluorimetry or immunological assay due to their wider biocompatibility, sensitivity, range limit, and precision detection. In this chapter, we summarize recent technological advances in melatonin quantification in plant samples and discuss the challenges of developed techniques in phytomelatonin research.

M. K. Hasan (✉)
Department of Agricultural Chemistry, Sylhet Agricultural University, Sylhet, Bangladesh
e-mail: hasanmk.agrichem@sau.ac.bd

J. Shopan
Key Laboratory of IRFA, Sylhet Agricultural University, Sylhet, Bangladesh

G. J. Ahammed
Department of Horticulture, Henan University of Science and Technology, Luoyang, Henan, China

J. Zhou (✉)
Zhejiang Provincial Key Laboratory of Horticultural Plant Integrative Biology, Department of Horticulture, Zijingang Campus, Zhejiang University, Hangzhou, China
e-mail: jie@zju.edu.cn

A. Sharma, G. J. Ahammed (eds.), *Melatonin in Plants: Role in Plant Growth, Development, and Stress Response*, Plant Life and Environment Dynamics,
https://doi.org/10.1007/978-981-99-8051-2_13

Keywords Phytomelatonin · Mass spectrophotometry · Electrochemical sensor · Fluorimetry · Nanodecoration

13.1 Introduction

Melatonin (*N*-acetyl-5-methoxytryptamine) as a protective phytohormone influentially regulates plant growth under various environmental circumstances (Arnao and Hernández-Ruiz 2019). Although, melatonin was initially identified as an animal hormone later the presence of melatonin in higher plants was confirmed, only 27 years ago in 1995 (Dubbels et al. 1995; Hattori et al. 1995). As an indolamine hormone melatonin is synthesized in plants from tryptophan. Phytomelatonin functions in diverse plant physiological actions. Additionally, melatonin draws an extra attention due to the modulator roles in sleep, body temperature regulation, mode, circadian rhythm, immunity, sexual behavior, and locomotors activity of humans (Bonomini et al. 2018; Arnao and Hernández-Ruiz 2018; Zhang et al. 2022). Though melatonin structure was discovered over 65 years ago in 1958, the progress of understanding its regulatory function in plant science is nonlinear and mostly governed by the development of new separation and detection techniques.

Melatonin is an ancient pleiotropic molecule present in algae to higher plants. This compound is able to cross the physiological barrier at the cellular level and reduce oxidative anxiety under stressed conditions. Reviews of the recent literature explored that melatonin is distributed throughout the plant system including root, stem, leaf, flower, fruits, and even seeds and oils but at differential gradients (Arnao and Hernández-Ruiz 2013). However, the highest amount of melatonin is obtained in nuts and germinated legumes and mushrooms, which reduces the appearance of many diseases upon consumption (Moser et al. 2022). Additionally, under stressed condition melatonin protects plants not only as a classical antioxidant but also stimulates other defense systems like regulation of the activity and expression of antioxidant and pro-oxidant enzymes similar to other plant hormones (Mercolini et al. 2012). Interestingly, the level of such a curious indolamine compound varies not only from species to species but also among the different varieties of the same plant species. Moreover, the examined tissue type, growth conditions, plant developmental stage, and various extraction methods and greatly the techniques used by scientists for detection highly affect the level of phytomelatonin content (Arnao and Hernández-Ruiz 2013; Hasan et al. 2015; Kanwar et al. 2020). Therefore a sensitive detection of melatonin in different plant organ and crop species is essential for future innovative research and human welfare (Khan et al. 2021).

At present, there are numerous available assay techniques used for the measurement of plant melatonin. On the basis of differentials interaction techniques with a mobile phase and a stationary phase the separation and quantification of plant melatonin by liquid chromatography (LC) is most common (Coskun 2016). Various investigations explored that among the LC technique high performance liquid chromatography (HPLC) equipped with different detectors including fluorescent detector (FD), UV/VIS photodiode array (PDA) detector or electrochemical detector

(ECD) are widely used. However, due to the high specificity and sensitivity with a low limit of detection (LOD), HPLC-FD with mass spectroscopy (MS) is the most popular (Afreen et al. 2006; Huang and Mazza 2011a). Similarly, very recently analysis of melatonin by HPLC-ESI-QTOF-MS equipped with RP-18 stationary column from plant matrix has drawn the attention by researchers due to their more precise and low limit of detection (Kukula-Koch et al. 2021). In contrast, the enzyme-linked immunosorbent assay (ELISA) or radioimmuno assay has gained popularity due to its reliable and efficient phytomelatonin detection, while high variability was observed in measurement compared to other detection techniques (Arnao and Hernández-Ruiz 2015a, b). Moreover, recently developed fast scan voltammetry (FSCV) in electrochemical detection and their nanofabrication most novel available technique used for melatonin detection (Castagnola et al. 2020). Again, as precision, simple and quick performing techniques spectrofluorimetry and chemiluminescence (CL) appeal an extra satisfaction of researchers worldwide. Although there are various melatonin detection tools and techniques, each of them has its own limitations or advantages. Therefore the choice of any detection tools depends on many factors such as specific research aims, availability of resources, and equipment with selectivity and sensitivity of the analytical techniques. Additionally, the main limitation of melatonin detection in plants is thought to be the lower concentration ranges from pico-gram to micro-gram per gram samples, which interns regulate the choice of sophisticated and highly sensitive tools and techniques in plant melatonin research (Cerezo et al. 2016). In this chapter, we abridged recent technological advances in phytomelatonin research and discussed the challenges of developed techniques.

13.2 Chromatographic Techniques

Based on differentials interaction techniques with a stationary phase and a mobile phase, the separation and identification of melatonin compound from a sample mixture has been widely used in various fields of investigation including plant science. Herein, a sample mixture is requisite to inject onto a stationary phase made of either liquid or solid materials and a mobile phase (liquid or gas) passes through the stationary phase. During the flows of the mobile phase, the sample mixture differentially interact with the stationary materials and is separated (Coskun 2016). Qualitative and quantitative measurement chromatographic identification and determination is an important biophysics technique. However, the analysis of phytomelatonin fronting has numerous challenges due to their low level of analytes and their quick reaction with reactive oxygen species (Huang and Mazza 2011a). Fascinatingly, scientists over the two decades have been working hard to find easy, quick, and sensitive separation and identification techniques of phytomelatonin analysis. There are several types of chromatographic techniques including liquid chromatography, liquid chromatography-tandem mass spectrometry, and capillary electrophoresis. Henceforth, we have discussed available chromatographic techniques in melatonin detection with their wide range of application in phytomelatonin research.

13.2.1 Liquid Chromatography (LC) Techniques

Among the chromatographic methods high performance liquid chromatography (HPLC) method is the most widely used melatonin separation and identification technique (Arnao and Hernández-Ruiz 2015a, b). In the HPLC system, a plant sample extract containing melatonin is injected into a solid phase stationary phase usually made of silica or a polymer allowing the separation of compounds based on their physico-chemical properties. The mobile phase, which is typically a mixture of solvents pushes the melatonin extract through the stationary phase or column; where melatonin is separated from other compounds due to their differential interaction with the stationary column (Huang and Mazza 2011a). After separation melatonin is quantified by using different detectors including the fluorescent detector (FD), UV/VIS photodiode array (PDA) detector or electrochemical detector (ECD) as schematically represented in Fig. 13.1. Therefore there is a large variation in sample melatonin analysis that may vary from the choice of the solvent mixture to the selection of stationary column and detector (Rzepka-Migut and Paprocka 2020). However, the use of this sophisticated technique for melatonin identification combined with measuring levels by means of liquid chromatography with electrochemical or fluorescence detection seems to be an efficient methodological option. Interestingly, most of the HPLC methods reviewed have used reverse phase columns (e.g., RP18 or RP8) for melatonin separation and fluorescence detectors that were found to be sensitive and versatile to quantify in most samples at low limits of detection and quantification (Garcia-Parrilla et al. 2009). Hasan et al. showed that endogenous melatonin concentration of tomato plants is sensibly possible to detect

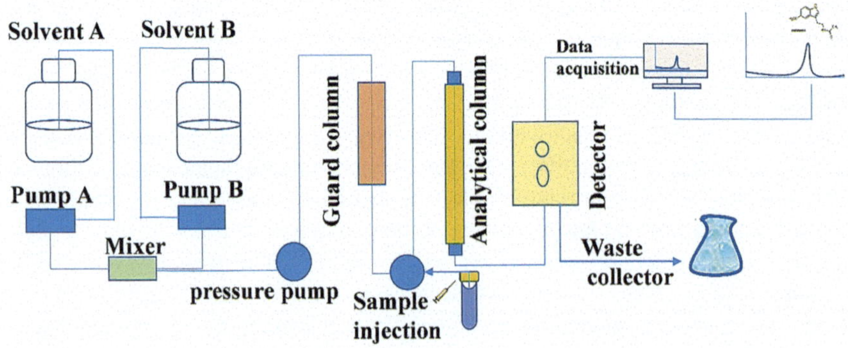

Fig. 13.1 Schematic diagram of a high performance liquid chromatography (HPLC) system used for the separation and identification of plant melatonin

by HPLC after solid phase extraction (SPE) using fluorescence detectors maintaining 280 nm excitation and 350 nm emission wavelength (Hasan et al. 2015). Alike, melatonin in different organs (leaves, roots, fruits, and seed) of pepper (*Capsicum annuum* L) plants under different growth stages including germination, seedling, flowering, and harvesting stage was detected using HPLC-FD equipped with 5 μm, 150 mm × 4.6 mm ODS-2 stationary column (Korkmaz et al. 2014). Interestingly in both cases, they observed very sensible separation and detection of melatonin from 10.1 ng- >250 ng/g fresh weight by using a mobile phase constituted with solvent mixture methanol and 0.1 mM Na_2HPO_4/H_3PO_4 acidic buffer at a flow rate 0.4 mL/min. Most of the published HPLC-FD methods used isocratic, linear or step gradient elution to separate melatonin. Interestingly, surfing literature, it has been observed that in all the cases either detected melatonin or serotonin not simultaneously together might be due to the similarity matrix of these two compounds.

In contrast, although the HPLC-ECD is not well defined, it is successively used to separate and quantify melatonin and serotonin in many plants. The first time HPLC-ECD was used to detect melatonin in brown algae was in 1996, afterward its presence was identified at a pico-gram level in many plant edible seeds, grass, walnut, and cherries using differentials gradient isocratic elution and mostly C18 reverse phased (RP) column or stationary phase (Huang and Mazza 2011a). However, the main interference observed is that similarity of a variety of compounds exist in plant tissues and their comparable oxidation potentials and similar retention times of melatonin or serotonin during separations (Huang and Mazza 2011a). Again, to determine the melatonin in the seed, leaf, stem, and roots of *Glycyrrhiza uralensis* F., Afreen et al. used HPLC- UV/VIS photodiode array (PDA) detector at the pico-gram level (Afreen et al. 2006). Interestingly, in HPLC- UV method melatonin and serotonin are simultaneously possible to separate and detect, however, the detection limit is higher in both the HPLC- UV or HPLC-ECD methods compared to HPLC-FD method (Huang and Mazza 2011a). Moreover, in HPLC-ECD the concentration of organic solvent in the mobile phase is usually higher (20%) which reduced the life span of the electrode in the detector. Therefore, the HPLC-FD method offers the best and most sensible separation and detection of melatonin in plants for routine measurement among the above-discussed different LC detectors (Huang and Mazza 2011b).

13.2.2 Liquid Chromatography Mass Spectrometry (LC-MS) Techniques

For avoiding interferences of other plant metabolites, separation, identification, and quantification of phytomelatonin from plant tissue by LC coupled with mass spectrometry (LC-MS) ranks prime choice by plant scientists as it is more precise and require no derivatization of sample extract due to selective ion-monitoring technique (Huang and Mazza 2011a). In this respect, studies published by Cao et al. (2006), where a robust method for determining melatonin, serotonin, and auxin in plant samples using LC-MS/MS clearly contribute to improving accurate research into melatonin in plants. Afterward, Huang and Mazza (2011b) did excellent work in the simultaneous measurement of serotonin and melatonin from fruits by using HPLC-MS equipped with electrospray ionization (ESI) in both positive and negative ionization modes. Interestingly, they achieved optimal analytical separation of melatonin and serotonin in the linear gradient elution of a mobile phase containing acetonitrile and water with 0.1% formic acid (Huang and Mazza 2011b). In fact, rapid advances in mass spectrometry have greatly attributed to the development progress of LC-MS including atmospheric-pressure ionization-mediated liquid introduction in the chromatography system. Interestingly, atmospheric-pressure solid analysis probe (ASAP) assisted mass spectrometry takes only 1 min for one single analysis of melatonin as no chromatographic separation is needed (Moser et al. 2022). Similarly, very recently HPLC-ESI-QTOF-MS equipped with RP-18 stationary column successfully separated and analyzed melatonin from plant matrix and the generated fingerprint varied in terms of signal intensity based on the quantity in sample extract (Kukula-Koch et al. 2021). Using LC-MS-ESI-single quadrupole detector, melatonin and serotonin in eight different strawberry cultivars were simultaneously detected (González-Gómez et al. 2009). In purple wheat, melatonin and serotonin were simultaneously measured through ultra-performance liquid chromatography mass spectrometry (UPLC-MS) using a binary gradient mobile phase and those two compounds were detected in a negative mode of ionization (Hosseinian et al. 2008). Although, the positive mode of ionization is more sensitive and common, melatonin and serotonin both the compounds have two nitrogen atoms in their structures that are easy to protonation (Huang and Mazza 2011a). However, compared to the LC the flow rate of the mobile phase in LC-MS is lower and the pH needs to be adjusted for high signal in the advanced MS.

Although LC-MS can separate and quantify melatonin directly, the GS-MS measured melatonin after the conversion of sample extract to a volatile derivative. For example, Vafadar et al. quantified melatonin in *Dracocephalum kotschyi* under salt stress by GC-MS, equipped with an HP-5MS capillary column (30 m × 0.25 mm I.D with 0.25 μM film thickness) maintaining 300 °C detector and 250 °C injector temperature, 13 psi pressure and helium as carrier gas (Vafadar et al. 2020). Similarly, under metal-stressed conditions Li et al. very sensibly and accurately measured the tomato plant endogenous melatonin content. Again, very recently, it has been observed that two-way micro fluidic splitter GC allowed simultaneous data acquisitions equipped with two different MS and FID detectors (Bahcesular et al. 2020). However, the main obstacle of LC-MS is the reduction of ionization efficiency that ultimately changes the peak area of analytes because of the co-eluting sample compound called the matrix effect. Therefore, though, liquid chromatography-tandem mass spectrometry (LC-MS/MS) techniques have aroused their enormity in plant melatonin research, these techniques are not infallible because of the matrix effect.

13.2.3 Capillary Electrophoresis (CE)

As a powerful chromatographic technique capillary electrophoresis (CE) separates melatonin from plant extract based on its charge and size using a high voltage electric field as passes through a separation channel filled with a buffer solution (Zhang et al. 2022). In recent years, it has been observed that the CE technique took place among the various established techniques of melatonin separation and detection in different fields of research including crop science (Przybylska et al. 2021). For example, in green tea (*Camellia sinensis*) low level of melatonin could be detected by the CE technique, where researchers showed that using a 10 mM/L background electrolyte (BGE) and 10% acetonitrile with sodium dodecyl sulfate (SDS) anionic micelles and β-CD (cyclodextrins) very quick successfully identified the antioxidant melatonin within 10 min (Gomez et al. 2015; Chrustek and Olszewska-Słonina 2021). Similarly, natural deep eutectic solvents (NADES) modified BGE showed excellent melatonin separation performance from plant materials (Zhang et al. 2022). Here, the choice of stationary phase plays crucial role in melatonin separation potentiality, hence scientists developed carboxyl multiwalled carbon nanotubes (c-MWNTs) immobilized capillary to detect phytomelatonin rapidly (Zhang et al. 2022). However, the recognition of CE in phytomelatonin is in increasing trend due to their high maturity and versatility of separation from small inorganic ions to high molecular–weight biomolecules (Voeten et al. 2018). Under optimal conditions, low limit of melatonin detection in different plant extracts aroused CE as an important contributor technique of phytomelatonin investigation.

13.3 Spectrofluorimetry

In fluorescence spectroscopy, also known as spectrofluorimetry or fluorimetry, the fluorescence intensity of a sample compound is measured after exciting at a particular wavelength. Using spectrofluorimetry assay, Oladi et al. satisfactorily detected higher amount of melatonin in pistachio nuts about 230 μg/g compared to the GC-MS and got ±14.3 pooled standard deviation among the replicates (Oladi et al. 2014). The excitation wavelength was maintained at 275 nm wavelength and the fluorescence emission intensity of sample melatonin was measured at 366 nm wavelength, where the author used ultrasound-assisted solid-liquid extraction to achieve the highest recovery. Due to the precision, simple and quick performing technique spectrofluorimetry appeal to the extra satisfaction of researchers. Again, the optical fluorimetric techniques attributed to the measurement of low cost, very high sensitivity, and extensive analyte concentration range (Vijayakumar et al. 2023). However, the accuracy of the measurement highly depends on a number of factors such as extraction efficiency, stability of extracted samples melatonin and spectrofluorimeter conditions during measurements. Therefore, further analysis is essential to confirm the identity of analytes by using other techniques such as

LC-MS. For instance, Suarez-Fernandez et al. detected root exudate metabolites by the spectrofluorimetric tool but they used HPLC for root tissue phytomelatonin determination (Suarez-Fernandez et al. 2020). Recently, Vijayakumar et al. developed antipyrine derivatives (APA) probe fluorescent technique toward melatonin determination in biofluids, where they use many other biologically important competitors such as proline, cysteine, tryptophan, dopamine ascorbic acid, glucose, and glutamine to observe the emission spectral interference (Vijayakumar et al. 2023). However, in phytomelatonin detection such kind of literature is still missing as the other indolamine compounds in plant samples show intrinsic fluorescence properties (Canizo et al. 2022). Over conventional techniques, Canizo et al. developed natural deep eutectic solvents (NADES) based fructose-citric acid comprehensive fluorescence technique for melatonin detection in biological samples and confirmed that solvent variability strongly affects the melatonin fluorescence intensity, nonetheless, further study is essential in the different eutectic systems (Canizo et al. 2022).

13.4 Immunological Assay

Among the several analytical techniques one of the most widely used techniques immunological assay of melatonin detection in fruits or edible plants because of their high specificity and sensitive detection techniques of target compounds (Gonçalves et al. 2021). Researchers rely on immunological assay of melatonin detection due to the specific binding potential of antibodies with melatonin or target compounds (Kennaway 2020), therefore high accuracy or sensitive detection of phytomelatonin is possible even in complex biological samples such as plant tissues or extracts and in various food products. There are two different types of immunological assay one is enzyme-linked immunosorbent assay (ELISA) and the other one is radioimmunoassay (RIA) generally used in plant melatonin determination.

13.4.1 Enzyme-Linked Immunosorbent Assay (ELISA)

For ELISA the main principle is based on the antigen–antibody interactions, where a specific antibody binds to the target melatonin (antigen) and forms a melatonin-antibody complex. These complexes are later detected by using another antibody called a secondary antibody that conjugates to an enzyme like horseradish peroxide (HRP) and gives a calorimetric reaction measured by using spectrophotometry (Kennaway 2020). For example, after the extraction of melatonin from plant tissues (cassava) with extraction buffer containing acetone, methanol and water at a ratio of 89:10:1 melatonin content determined by using a competitive enzyme immunoassay kit purchased from Haling Biotechnology (cat-HLE97243), where an experimental sample with standard load loaded to the template and added horseradish peroxide (HRP)-conjugates reagent before incubation at 37 °C for 1 h (Bai et al. 2022). After washing with buffer and chromatographic reaction optical density was measured at 450 nm wavelength in microplate reader (Bai et al. 2022). In progress, Chen and his co-worker performed excellent research to improve the immunological detection technique for phytomelatonin determination in different food samples. For example, in 2017, they developed a rabbit polyclonal antibody-based a completive indirect enzyme-linked immune-sorbent assay (ci-ELISA) (Chen et al. 2017). To further improvement in shortening the detection time they a colloidal gold immunochromatograpy (ICA) strip test method was developed for successful food melatonin determination. However, the development of monoclonal antibody labeled with colloidal gold nanoparticles is the main problem immunochromatographic assay (Chen et al. 2019). Interestingly, Johns et al. reported that the ELISA technique creates less cross-reactivity with other compounds of fruits and does not affect the melatonin level that co-eluted during HPLC separation (Johns et al. 2013; Feng et al. 2014). Although ELISA provides reliable and efficient techniques for phytomelatonin detection while very recently, Peña-Delgado et al. ram seminal plasma melatonin detected by using a commercial competitive immunoassay (ELISA) kit though they used HPLC-ESI-MS/MS triple quadrupole for melatonin determination in vegetables and agri-food by-products (Peña-Delgado et al. 2023). These may be due to the avoidance interference of other plant compounds of immunoassay.

13.4.2 Radioimmunoassay (RIA)

To perform RIA in plant melatonin detection an antibody specific to the melatonin labeled with a radioactive isotope like iodine-125 or iodine-131 using a specific reagent such as iodobeads or chloramine-T and form an antigen-antibody complex (Rzepka-Migut and Paprocka 2020). Afterward, the measured amount of radioactivity of the formed complex is considered as the concentration of targeted melatonin in plant tissues. Interestingly, it has been previously reported that quantifying the melatonin amount in tomato fruit was higher than the amount quantified by GS-MS (Van Tassel and O'Neill 2001). These may be due to the cross-react interference of other plant molecules with related antibodies and enzymes that potentially leads to the false reading or over-estimation of the actual amount (Feng et al. 2014). Although the RIA technique has been widely used in phytomelatonin detection in many plant species including rice, banana, grapes, tomato, beetroot cucumber, and tobacco lease, however, it is less specific compared to the LC technique (Grao-Cruces et al. 2023; Dubbels et al. 1995; Reiter et al. 2015). Recently it has been reviewed that there are many companies supplying immunoassay kits that need comparative validation of sensitivity and specificity in phytomelatonin determination (Kennaway 2020). Moreover, such technology needs very well trained expertise personnel to operate the sophisticated specialized equipment. Unlike animals, the immunological RIA or ELISA techniques create serious problems due to the cross-reactivity with other co-extractives indoleamine (Arnao and Hernández-Ruiz

2015a, b). Moreover, melatonin is not an immunogenic compound, results in the antibody used in immunology must need to link with an immunogenic compound that strongly influences the variability of quality assessment of phytomelatonin (Arnao and Hernández-Ruiz 2015a, b).

13.5 Chemiluminescence (CL)

In chemiluminescence techniques of melatonin determination, a light emission occurred due to a specific chemical reaction which is then measured to determine the melatonin concentration in plant tissues. Generally, during the light emission a chemical reaction occurred between phytomelatonin and hydrogen peroxide in the presence of horseradish peroxide catalyst, where highly reactive peroxyl radical generate that reacts with a luminescent compound like luminol or derivatives and produce light in the visible spectral range (Sariahmetoglu et al. 2003; Liu et al. 2016). The intensity of emitted light is considered proportional to phytomelatonin concentration. The chemiluminescence technique of melatonin determination is very specific and highly sensitive and easy in compared to LC or ELISA techniques. However, the chemiluminescence was reduced or inhibited due to the interference of another indoleamine in the sample extract (Wei et al. 2007). Based on ultra-weak chemiluminesence generated by melatonin under the alkaline condition in fruit and vegetable samples measured by Lu et al. (2002), they get low sensitivity and need completely pure samples to avoid interference with other indoleamine. In contrast, the electro chemiluminescence (ECL) technique attracts a lot of interest because of its first response, sensitivity, selectivity, and simplified instrumentation, high energy electron transfer reactions, wide linear response range (1.0×10^{-14} to 1.0×10^{-9} M), excellent detection limit (6.2 fM), and reproducibility (Hosseini et al. 2022). Interestingly, recent year's integration of chemiluminescence with capillary electrophoresis chromatographic technique draw attention to biomolecules analysis due to their distinct advantages including the reduced sample or solvent consumption extended resolution and low level rapid determination, wider perspective phytomelatonin research (Liu et al. 2016).

13.6 Electrochemical Methods

Over the 8 years (2015 onwards) the development of novel sensor-based electrochemical or optical detection of melatonin aroused as a new platform of research, where scientists intended to decorate the sensitivity and selectivity of techniques through surface modification or fabrication by using diverse nanomaterials bearing unique properties of melatonin or tryptophan detection (Khan et al. 2021). In recent years, electrochemical methods rises as promising tools due to their simplicity in instrumentation, ease to use, rapid and low cost sensitive detection at a low limit

(LOD) in both the in vivo and real-time sensing and portability advantage (Khan et al. 2021; Castagnola et al. 2020). In electrochemical detection minimum two electrodes consist where one functioned as working electrode and the other is a counter or auxiliary electrode. These electrode contact each other either by an electrically conductive media or an external electric circuit (Scozzari 2008; Khan et al. 2021). In comparison to conventional LC-MS techniques or immunological assay their several types of electrochemical techniques such as square wave pulse voltammetry (SWV), fast scan cyclic voltammetry (FSCV) and their nano-fabrications (Castagnola et al. 2020; Hensley et al. 2018; Lete et al. 2022). In recent years, the electrochemical sensor-based carbon electrode attracted great interest due to their high sensitivity, chemical inertness, low cost, and easy decoration. Hence, as a cost-effective eco-friendly technique carbon paste electrode, reduced grapheme oxide electrode, FeCo nanofiber, palladium nanocomposite and multiwalled carbon nanotube (MWCN)-dihexadecyl hydrogen phosphate embedded electrode among many other has been widely investigated (Lete et al. 2022; Zeinali et al. 2017; Duan et al. 2020; Kumar and Goyal 2016). Consequently, Lete et al. very recently developed an electrochemical sensor-based sonogel-carbon electrode (SNGCE) that enriched with gold nanoparticle (AuNP) and the performance validated with cyclic voltammetry in the presence of a redox probe potassium hexacyanoferrate (Lete et al. 2022). However, the literature relating to their application in plant or foodstuff melatonin detection remains elusive. In connection, Afshar et al. developed an ultrasensitive and extremely selective 'turn-on' fluorescence sensor for the measurement of melatonin in many juice samples (Afshar et al. 2022). To achieve the reliability and sensitivity of the measurement they used Zr metal-organic framework encapsulation into the molecularly imprinted polymer fluorescence probe (MOF@MIP). Interestingly, the developed method was able to detect a trace amount of melatonin (0.18 ng/mL) in grape, cherry, and sour cherry juice, which suggested the enormous potential of MOF@MIP fabricated sensor for juice samples analysis (Afshar et al. 2022). Though, the presence of other interfering compounds can influence the electrometrical detections by affecting the selectivity of sensors. Therefore an in-depth research investigation is essential to confirm the fabricated technology and provide a platform for melatonin detection in other plant and food samples.

13.7 Conclusions and Future Perspectives

As an ancient molecule melatonin is present in algae to higher plants, fruits, edible seeds, juices, and varieties of other food that has broader application from plants to animals or humans due to its multitude of physiological functions (Zhang et al. 2022). Therefore, a sensitive, reliable easy and quick analysis in various plants, fruits or foodstuff is essential to compile data information for a better understanding of the significance of the phytomelatonin. Interestingly, due to the high specificity of HPLC-FD for indoleamines with their substitution and electron delocalization at specific absorption or emission wave length, this technique of melatonin detection was validated as reliable and met the requirements of the association of official agricultural chemistry (AOAC) compared to the LC-MS (Johns et al. 2013). Therefore, many researchers also chose HPLC-FD instead of LC-MS or both for the quantitative measurement of melatonin in fruit or plant samples. Again, melatonin is not immunogenic so the antibody produced for the immunoassay (especially ELISA and RIA technology) must need chemically linked to the immunogenic melatonin compound that regulates variability of quality assessment in different samples due to cross-reactivity with co-extractives (Kennaway 2020; Arnao and Hernández-Ruiz 2015a, b). Similarly, the updates of electrochemical or optical sensor-based detection by using different electrode materials and their necessary fabrication are expected to strengthen the technological advances in phytomelatonin research. Similarly, the recent addition of electro chemiluminescence (ECL) technique or integration of chemiluminescence with capillary electrophoresis chromatographic technique is attracting a lot of interest because of its first response, sensitivity, selectivity, wide linear response range, high reproducibility, and excellent detection limit. Again, nanodecoration of novel electrochemical or optical sensor-based detection tools aroused their enormity in phytomelatonin research due to their ability to offer high specificity, ultra-sensitivity, better signal-to-noise ratio, and wide dynamic range (Khan et al. 2021). However, a more in-depth research is essential for technological advancement and to unravel much more unknown physiological relevance of phytomelatonin to human welfare.

Acknowledgments This work was supported by the National Science and Technology (NST) of Bangladesh and Sylhet Agricultural university research System (SAURES), Sylhet-3100, Bangladesh.

Author Contributions M.K.H and J.S planned and wrote initial draft of the manuscript. M.K.H., G.JA. and J.Z revised the manuscript.

References

Afreen F, Zobayed SM, Kozai T (2006) Melatonin in Glycyrrhiza uralensis: response of plant roots to spectral quality of light and UV-B radiation. J Pineal Res 41(2):108–115

Afshar EA, Taher MA, Karimi F, Karaman C, Moradi O (2022) Ultrasensitive and highly selective "turn-on" fluorescent sensor for the detection and measurement of melatonin in juice samples. Chemosphere 295:133869

Arnao MB, Hernández-Ruiz J (2013) Growth conditions influence the melatonin content of tomato plants. Food Chem 138(2–3):1212–1214

Arnao MB, Hernández-Ruiz J (2015a) Chapter 11. Phytomelatonin: searching for plants with high levels for use as a natural nutraceutical. In: Atta ur R (ed) Studies in natural products chemistry, vol 46. Elsevier, Amsterdam, pp 519–545. https://doi.org/10.1016/B978-0-444-63462-7.00011-7

Arnao MB, Hernández-Ruiz J (2015b) Functions of melatonin in plants: a review. J Pineal Res 59(2):133–150

Arnao MB, Hernández-Ruiz J (2018) Phytomelatonin, natural melatonin from plants as a novel dietary supplement: sources, activities and world market. J Funct Foods 48:37–42

Arnao MB, Hernández-Ruiz J (2019) Melatonin: a new plant hormone and/or a plant master regulator? Trends Plant Sci 24(1):38–48

Bahcesular B, Yildirim ED, Karaçocuk M, Kulak M, Karaman S (2020) Seed priming with melatonin effects on growth, essential oil compounds and antioxidant activity of basil (*Ocimum basilicum* L.) under salinity stress. Ind Crop Prod 146:112165

Bai Y, Wei Y, Yin H, Hu W, Cheng X, Guo J, Dong Y, Zheng L, Xie H, Zeng H (2022) PP2C1 fine-tunes melatonin biosynthesis and phytomelatonin receptor PMTR1 binding to melatonin in cassava. J Pineal Res 73(1):e12804

Bonomini F, Borsani E, Favero G, Rodella LF, Rezzani R (2018) Dietary melatonin supplementation could be a promising preventing/therapeutic approach for a variety of liver diseases. Nutrients 10(9):1135

Canizo BV, Quintas PY, Wuilloud RG, Silva MF, Gomez FJ (2022) Fluorescent behavior of melatonin and related indoleamines in natural deep eutectic solvents. J Mol Liq 363:119902

Cao J, Murch SJ, O'Brien R, Saxena PK (2006) Rapid method for accurate analysis of melatonin, serotonin and auxin in plant samples using liquid chromatography–tandem mass spectrometry. J Chromatogr A 1134(1–2):333–337

Castagnola E, Woeppel K, Golabchi A, McGuier M, Chodapaneedi N, Metro J, Taylor IM, Cui XT (2020) Electrochemical detection of exogenously administered melatonin in the brain. Analyst 145(7):2612–2620

Cerezo AB, Leal Á, Álvarez-Fernández MA, Hornedo-Ortega R, Troncoso AM, García-Parrilla MC (2016) Quality control and determination of melatonin in food supplements. J Food Compos Anal 45:80–86

Chen Z-J, Fu H-J, Luo L, Sun Y-M, Yang J-Y, Zeng D-P, Shen Y-D, Xu Z-L (2017) Development of competitive indirect ELISAs with a flexible working range for the simple quantification of melatonin in medicinal foods. Anal Methods 9(10):1617–1626

Chen Z, Zhou K, Ha W, Chen P, Fu H, Shen Y, Sun Y, Xu Z (2019) Development of a low-cost, simple, fast and quantitative lateral-flow immunochromatographic assay (ICA) strip for melatonin in health foods. Food Agric Immunol 30(1):497–509

Chrustek A, Olszewska-Słonina D (2021) Melatonin as a powerful antioxidant. Acta Pharm 71(3): 335–354

Coskun O (2016) Separation techniques: chromatography. North Clin Istanbul 3(2):156

Duan D, Ding Y, Li L, Ma G (2020) Rapid quantitative detection of melatonin by electrochemical sensor based on carbon nanofibers embedded with FeCo alloy nanoparticles. J Electroanal Chem 873:114422

Dubbels R, Reiter R, Klenke E, Goebel A, Schnakenberg E, Ehlers C, Schiwara H, Schloot W (1995) Melatonin in edible plants identified by radioimmunoassay and by high performance liquid chromatography-mass spectrometry. J Pineal Res 18(1):28–31

Feng X, Wang M, Zhao Y, Han P, Dai Y (2014) Melatonin from different fruit sources, functional roles, and analytical methods. Trends Food Sci Technol 37(1):21–31

Garcia-Parrilla MC, Cantos E, Troncoso AM (2009) Analysis of melatonin in foods. J Food Compos Anal 22(3):177–183

Gomez FJ, Hernández IG, Cerutti S, Silva MF (2015) Solid phase extraction/cyclodextrin-modified micellar electrokinetic chromatography for the analysis of melatonin and related indole compounds in plants. Microchem J 123:22–27

Gonçalves AC, Nunes AR, Alves G, Silva LR (2021) Serotonin and melatonin: plant sources, analytical methods, and human health benefits. Rev Bras 31:162–175

González-Gómez D, Lozano M, Fernández-León MF, Ayuso M, Bernalte MJ, Rodríguez AB (2009) Detection and quantification of melatonin and serotonin in eight sweet cherry cultivars (*Prunus avium* L.). Eur Food Res Technol 229:223–229

Grao-Cruces E, Calvo JR, Maldonado-Aibar MD, Millan-Linares MC, Montserrat-de la Paz S (2023) Mediterranean diet and melatonin: a systematic review. Antioxidants 12(2):264

Hasan MK, Ahammed GJ, Yin L, Shi K, Xia X, Zhou Y, Yu J, Zhou J (2015) Melatonin mitigates cadmium phytotoxicity through modulation of phytochelatins biosynthesis, vacuolar sequestration, and antioxidant potential in *Solanum lycopersicum* L. Front Plant Sci 6:601

Hattori A, Migitaka H, Iigo M, Itoh M, Yamamoto K, Ohtani-Kaneko R, Hara M, Suzuki T, Reiter RJ (1995) Identification of melatonin in plants and its effects on plasma melatonin levels and binding to melatonin receptors in vertebrates. Biochem Mol Biol Int 35(3):627–634

Hensley AL, Colley AR, Ross AE (2018) Real-time detection of melatonin using fast-scan cyclic voltammetry. Anal Chem 90(14):8642–8650

Hosseini M, Hashemian E, Salehnia F, Ganjali MR (2022) Turn-on electrochemiluminescence sensing of melatonin based on graphitic carbon nitride nanosheets. J Electroanal Chem 921: 116593

Hosseinian FS, Li W, Beta T (2008) Measurement of anthocyanins and other phytochemicals in purple wheat. Food Chem 109(4):916–924

Huang X, Mazza G (2011a) Application of LC and LC-MS to the analysis of melatonin and serotonin in edible plants. Crit Rev Food Sci Nutr 51(4):269–284

Huang X, Mazza G (2011b) Simultaneous analysis of serotonin, melatonin, piceid and resveratrol in fruits using liquid chromatography tandem mass spectrometry. J Chromatogr A 1218(25): 3890–3899

Johns NP, Johns J, Porasuphatana S, Plaimee P, Sae-Teaw M (2013) Dietary intake of melatonin from tropical fruit altered urinary excretion of 6-sulfatoxymelatonin in healthy volunteers. J Agric Food Chem 61(4):913–919

Kanwar MK, Xie D, Yang C, Ahammed GJ, Qi Z, Hasan MK, Reiter RJ, Yu J-Q, Zhou J (2020) Melatonin promotes metabolism of bisphenol a by enhancing glutathione-dependent detoxification in *Solanum lycopersicum* L. J Hazard Mater 388:121727

Kennaway DJ (2020) Measuring melatonin by immunoassay. J Pineal Res 69(1):e12657

Khan ZA, Hong PJ-S, Lee CH, Hong Y (2021) Recent advances in electrochemical and optical sensors for detecting tryptophan and melatonin. Int J Nanomedicine 16:6861

Korkmaz A, Değer Ö, Cuci Y (2014) Profiling the melatonin content in organs of the pepper plant during different growth stages. Sci Hortic 172:242–247. https://doi.org/10.1016/j.scienta.2014.04.018

Kukula-Koch W, Szwajgier D, Gaweł-Bęben K, Strzępek-Gomółka M, Głowniak K, Meissner HO (2021) Is Phytomelatonin complex better than synthetic melatonin? The assessment of the antiradical and anti-inflammatory properties. Molecules 26(19):6087

Kumar N, Goyal RN (2016) Nanopalladium grained polymer nanocomposite based sensor for the sensitive determination of melatonin. Electrochim Acta 211:18–26

Lete C, López-Iglesias D, García-Guzmán JJ, Leau S-A, Stanciu AE, Marin M, Palacios-Santander JM, Lupu S, Cubillana-Aguilera L (2022) A sensitive electrochemical sensor based on sonogel-carbon material enriched with gold nanoparticles for melatonin determination. Sensors 22(1): 120

Liu M, Wang Z-Y, Zhang C-Y (2016) Recent advance in chemiluminescence assay and its biochemical applications. Chin J Anal Chem 44(12):1934–1941. https://doi.org/10.1016/S1872-2040(16)60981-7

Lu J, Lau C, Lee MK, Kai M (2002) Simple and convenient chemiluminescence method for the determination of melatonin. Anal Chim Acta 455(2):193–198

Mercolini L, Mandrioli R, Raggi MA (2012) Content of melatonin and other antioxidants in grape-related foodstuffs: measurement using a MEPS-HPLC-F method. J Pineal Res 53(1):21–28

Moser D, Hussain S, Rainer M, Jakschitz T, Bonn GK (2022) A validated method for the rapid quantification of melatonin in over-the-counter hypnotics by the atmospheric pressure solid analysis probe (ASAP). Anal Methods 14(16):1603–1610

Oladi E, Mohamadi M, Shamspur T, Mostafavi A (2014) Spectrofluorimetric determination of melatonin in kernels of four different Pistacia varieties after ultrasound-assisted solid–liquid extraction. Spectrochim Acta A Mol Biomol Spectrosc 132:326–329

Peña-Delgado V, Carvajal-Serna M, Fondevila M, Martín-Cabrejas MA, Aguilera Y, Álvarez-Rivera G, Abecia JA, Casao A, Pérez-Pe R (2023) Improvement of the seminal characteristics in rams using agri-food by-products rich in phytomelatonin. Animals 13(5):905

Przybylska A, Gackowski M, Koba M (2021) Application of capillary electrophoresis to the analysis of bioactive compounds in herbal raw materials. Molecules 26(8):2135

Reiter RJ, Tan D-X, Zhou Z, Cruz MHC, Fuentes-Broto L, Galano A (2015) Phytomelatonin: assisting plants to survive and thrive. Molecules 20(4):7396–7437

Rzepka-Migut B, Paprocka J (2020) Melatonin-measurement methods and the factors modifying the results. A systematic review of the literature. Int J Environ Res Public Health 17(6):1916

Sariahmetoglu M, Wheatley RA, Çakýcý Ý, Kanzýk Ý, Townshend A (2003) Evaluation of the antioxidant effect of melatonin by flow injection analysis-luminol chemiluminescence. Pharmacol Res 48(4):361–367. https://doi.org/10.1016/S1043-6618(03)00177-4

Scozzari A (2008) Electrochemical sensing methods: a brief review. In: Algal toxins: nature, occurrence, effect and detection. Springer Science + Business Media B.V., Dordrecht, pp 335–351

Suarez-Fernandez M, Marhuenda-Egea FC, Lopez-Moya F, Arnao MB, Cabrera-Escribano F, Nueda MJ, Gunsé B, Lopez-Llorca LV (2020) Chitosan induces plant hormones and defenses in tomato root exudates. Front Plant Sci 11:572087

Vafadar F, Amooaghaie R, Ehsanzadeh P, Ghanati F, Sajedi RH (2020) Crosstalk between melatonin and Ca2+/CaM evokes systemic salt tolerance in Dracocephalum kotschyi. J Plant Physiol 252:153237

Van Tassel DL, O'Neill SD (2001) Putative regulatory molecules in plants: evaluating melatonin. J Pineal Res 31(1):1–7

Vijayakumar S, Venkatesan S, Lin M-C, Vediappen P (2023) Highly sensitive and selective detection of melatonin in biofluids by Antipyrine based fluorophore. J Fluoresc 33(1):383–392

Voeten RL, Ventouri IK, Haselberg R, Somsen GW (2018) Capillary electrophoresis: trends and recent advances. Anal Chem 90(3):1464–1481

Wei SL, Zhao LX, Cheng XL, Lin JM (2007) Determination of melatonin in rat pineal gland and drug with flow-injection chemiluminescence. Chin J Chem 25(4):535–541

Zeinali H, Bagheri H, Monsef-Khoshhesab Z, Khoshsafar H, Hajian A (2017) Nanomolar simultaneous determination of tryptophan and melatonin by a new ionic liquid carbon paste electrode modified with SnO2-Co3O4@ rGO nanocomposite. Mater Sci Eng C 71:386–394

Zhang X-Y, Zhang Y, Yu Z, Liu Z-f, Wei B-B, Feng X-S (2022) Melatonin in different food samples: recent update on distribution, bioactivities, pretreatment and analysis techniques. Food Res Int 163:112272

The manufacturer's authorised representative in the EU is Springer Nature Customer Service Centre GmbH, Europaplatz 3, 69115 Heidelberg, Germany. If you have any concerns regarding our products, please contact ProductSafety@springernature.com

Printed and bound by CPI Group (UK) Ltd, Croydon, CR0 4YY
01/10/2026
02225855-0002